Computational Biomechanics for Medicine

Karol Miller • Adam Wittek • Grand Joldes
Martyn P. Nash • Poul M. F. Nielsen

Editors

Computational Biomechanics for Medicine

Solid and Fluid Mechanics for the Benefit of Patients

 Springer

Editors
Karol Miller
Intelligent Systems for Medicine Laboratory
The University of Western Australia
Perth, WA, Australia

Grand Joldes
Intelligent Systems for Medicine Laboratory
The University of Western Australia
Perth, WA, Australia

Poul M. F. Nielsen
Department of Engineering Science
Auckland Bioengineering Institute
University of Auckland
Auckland, New Zealand

Adam Wittek
Intelligent Systems for Medicine Laboratory
The University of Western Australia
Perth, WA, Australia

Martyn P. Nash
Department of Engineering Science
Auckland Bioengineering Institute
University of Auckland
Auckland, New Zealand

ISBN 978-3-030-42430-5   ISBN 978-3-030-42428-2   (eBook)
https://doi.org/10.1007/978-3-030-42428-2

This Springer imprint is published by the registered company Springer Nature Switzerland AG.
The registered company address is: Gewerbestrasse 11, 6330 Cham, Switzerland

Preface

Extending the success of computational mechanics to fields outside traditional engineering, in particular to biology, the biomedical sciences and medicine has been recognised as one of the greatest challenges facing the computational engineering and computational mechanics communities. While advancements are being made towards clinically relevant computational biomechanics models and simulations, there is still much work ahead before personalised medicine underpinned by personalised computer simulations becomes an integral part of healthcare.

The first volume in the *Computational Biomechanics for Medicine* book series was published in 2010. Since then, the book has become an annual forum for specialists in computational sciences to describe their latest results and discuss the possibility of applying their techniques to computer-integrated medicine. This eleventh volume in the *Computational Biomechanics for Medicine* book series comprises eleven of the latest developments in continuum biomechanics and patient-specific computations, by researchers from Australia, New Zealand, China, France, Germany, Greece and Poland. Some of the topics covered in this book are as follows:

- Medical image analysis
- Image-guided surgery
- Surgical intervention planning
- Disease prognosis and diagnosis
- Cell biomechanics
- Soft tissue biomechanics
- Injury mechanism analysis

The *Computational Biomechanics for Medicine* book series does not only provide the community with a snapshot of the latest state of the art, but more importantly, when computational biomechanics and patient-specific modelling is a mainstay of personalised healthcare, it will serve as a key reminder of how the field has overcome one of its greatest challenges.

Perth, Australia   Karol Miller
Perth, Australia   Adam Wittek
Perth, Australia   Grand Joldes
Auckland, New Zealand   Martyn P. Nash
Auckland, New Zealand   Poul M. F. Nielsen

Contents

**What Has Image Based Modelling of Cerebrospinal Fluid Flow
in Chiari Malformation Taught Us About Syringomyelia
Mechanisms?** 1
Lynne E. Bilston

Part I   Computational Solid Mechanics

**Lung Tumor Tracking Based on Patient-Specific Biomechanical
Model of the Respiratory System** 5
Hamid Ladjal, Michael Beuve, and Behzad Shariat

Design of Auxetic Coronary Stents by Topology Optimization 17
Huipeng Xue and Zhen Luo

**Physics-Based Deep Neural Network for Real-Time Lesion
Tracking in Ultrasound-Guided Breast Biopsy** 33
Andrea Mendizabal, Eleonora Tagliabue, Jean-Nicolas Brunet,
Diego Dall'Alba, Paolo Fiorini, and Stéphane Cotin

**An Improved Coarse-Grained Model to Accurately Predict Red
Blood Cell Morphology and Deformability** 47
Nadeeshani Maheshika Geekiyanage, Robert Flower, Yuan Tong Gu,
and Emilie Sauret

**Development of a Computational Modelling Platform
for Patient-specific Treatment of Osteoporosis** 85
Madge Martin, Vittorio Sansalone, and Peter Pivonka

Part II   Topics in Patient-Specific Computations

**Towards Visualising and Understanding Patient-Specific
Biomechanics of Abdominal Aortic Aneurysms** 111
K. R. Beinart, George C. Bourantas, and Karol Miller

Pipeline for 3D Reconstruction of Lung Surfaces Using Intrinsic Features Under Pressure-Controlled Ventilation 123
Samuel Richardson, Thiranja P. Babarenda Gamage, Toby Jackson,
Amir HajiRassouliha, Alys Clark, Martyn P. Nash, Andrew Taberner,
Merryn H. Tawhai, and Poul M. F. Nielsen

A Flux-Conservative Finite Difference Scheme for Anisotropic Bioelectric Problems 135
George C. Bourantas, Benjamin F. Zwick, Simon K. Warfield,
Damon E. Hyde, Adam Wittek, and Karol Miller

A Fast Method of Virtual Stent Graft Deployment for Computer Assisted EVAR 147
Aymeric Pionteck, Baptiste Pierrat, Sébastien Gorges, Jean-Noël Albertini,
and Stéphane Avril

Efficient GPU-Based Numerical Simulation of Cryoablation of the Kidney 171
Joachim Georgii, Torben Pätz, Christian Rieder, Hanne Ballhausen,
Michael Schwenke, Lars Ole Schwen, Sabrina Haase, and Tobias Preusser

Index 195

What Has Image Based Modelling of Cerebrospinal Fluid Flow in Chiari Malformation Taught Us About Syringomyelia Mechanisms?

Lynne E. Bilston

Abstract Chiari Malformation is a congenital disorder of the hindbrain, in which the cerebellar tonsils protrude through the foramen magnum, impeding normal cerebrospinal fluid (CSF) flow into the spinal canal. It is associated with pain, dizziness and headaches, particularly related to coughing and straining. The mechanisms by which Chiari malformation gives rise to these symptoms are not understood. In a large proportion of patients, a fluid-filled cavity develops in the spinal cord, called a syrinx. Syrinxes can cause additional neurological deficits, including sensory changes, weakness and upper limb pain. Syrinxes are associated with disturbances to normal CSF dynamics, usually as a result of obstructions in the spinal canal, but precisely how this occurs is not known. Animal studies suggest that fluid transport into the spinal cord is increased in the presence of spinal canal obstructions, likely via annular spaces surrounding penetrating arteries (perivascular spaces). Human phase contrast magnetic resonance imaging studies can quantify both cardiac driven motion of cerebrospinal fluid flow, and, more recently, respiratory and other influences. These data can be used to generate subject-specific computational fluid dynamics models of the hindbrain and spinal canal to estimate spinal canal pressure dynamics in patients with Chiari malformation, patients with syrinxes, and healthy controls. Computational models of perivascular space flow can be linked to these macroscopic models, to enable investigation of the feasibility of hypotheses about mechanisms of syrinx formation. To date, these studies have demonstrated that several popular hypotheses about Chiari mechanisms and syrinx formation are inconsistent with the mechanics of CSF flow, and generated novel mechanistic hypotheses. Subject-specific image based modelling provide a useful adjunct to human and animal experimental research into CSF flow disorders such as Chiari malformation and syringomyelia.

L. E. Bilston (✉)
Neuroscience Research Australia and Prince of Wales Clinical School, UNSW, Sydney, NSW, Australia
e-mail: L.Bilston@neura.edu.au

© Springer Nature Switzerland AG 2020
K. Miller et al. (eds.), *Computational Biomechanics for Medicine*,
https://doi.org/10.1007/978-3-030-42428-2_1

Part I
Computational Solid Mechanics

Lung Tumor Tracking Based on Patient-Specific Biomechanical Model of the Respiratory System

Hamid Ladjal, Michael Beuve, and Behzad Shariat

Abstract In this chapter, we evaluate the 3D tumor trajectories from patient-specific biomechanical model of the respiratory system, which takes into account the physiology of respiratory motion to simulate irregular motion. The behaviour of the lungs, driving directly by simulated actions of the breathing muscles, i.e. the diaphragm and the intercostal muscles (the rib cage). In this chapter, the lung model is monitored and controlled by a personalized lung pressure-volume relationship during a whole respiratory cycle. The lung pressure is patient specific and calculated by an optimization framework based on inverse finite element analysis. We have evaluated the motion estimation accuracy on two selected patients, with small and large breathing amplitudes (Patient $1 = 10.9$ mm, Patient $10 = 26.06$ mm). In this order, the lung tumor trajectories identified from 4D CT scan images were used as reference and compared with the 3D lung tumor trajectories estimated from finite element simulation during the whole cycle of breathing. Over all phases of respiration, the average mean error is less than 1.8 ± 1.3 mm. We believe that this model, despite of others takes into account the challenging problem of the respiratory variabilities and can potentially be incorporated effectively in Treatment Planning System (TPS) and as lung tumor motion tracking system during radiation treatment.

Keywords Biomechanics · Respiratory motion · Breathing mechanics · Lung tumor tracking · Radiation therapy · Medical imaging · Finite element method

H. Ladjal (✉)
LIRIS, Laboratoire d'InfoRmatique en Image et Systèmes d'information CNRS UMR 5205, University of Lyon, Université Claude Bernard Lyon 1, Lyon, France

IP2I CNRS UMR 5822, University of Lyon, Université Claude Bernard Lyon 1, Lyon, France
e-mail: hamid.ladjal@liris.cnrs.fr; hamid.ladjal@univ-lyon1.fr

M. Beuve
IP2I CNRS UMR 5822, University of Lyon, Université Claude Bernard Lyon 1, Lyon, France

B. Shariat
LIRIS, Laboratoire d'InfoRmatique en Image et Systèmes d'information CNRS UMR 5205, University of Lyon, Université Claude Bernard Lyon 1, Lyon, France

© Springer Nature Switzerland AG 2020
K. Miller et al. (eds.), *Computational Biomechanics for Medicine*,
https://doi.org/10.1007/978-3-030-42428-2_2

1 Introduction

Organ motion due to patient breathing introduces a technical challenge for dosimetry and lung tumor treatment by radiation therapy. Accurate dose distribution estimation requires patient-specific information on tumor position, size and shape as well as information regarding the material density and stopping power of the media along the beam path. In order to calculate and to ensure sufficient dose coverage throughout the treatment, the internal margin (IM) and setup margin (SM) are added to the clinical target volume (CTV) to compensate for the breathing movement and to obtain target volume (PTV). Generally, the addition of different margins leads to an excessively large PTV that would go beyond the patient's tolerance, and does not reflect the actual clinical consequences [1]. In the case of moving tumors, the PTV is increased so that the tumor lies inside the treatment field at all times. Breathing is an active and a complex process where the respiratory motion is non-reproducible, and the breathing periodicity, amplitude and motion path of patients' organs are observed during the respiration [2, 3]. Various different types of correspondence models that have been used and developed in the literature (linear, piece-wise linear, polynomial, B-spline, neural networks) in order to correlate the internal motion to respiratory surrogate signals. For more information on the correspondence models please see the complete review in Chap. III of Ehrhardt & Lorenz 2013 [3].

The biomechanical approaches aim at identification and taking into account the different anatomical and physiological aspects of breathing dynamics. These approaches attempt to describe respiratory-induced organ motion through a mathematical formulation based on continuum media mechanics solved generally on Finite Element Methods (FEM) [4–6]. Unfortunately, most of the time, the authors have used a single organ (lung) with nonrealistic of boundary conditions, or the lung motion is simulated by using simple displacement boundary conditions which are not realistic and do not take into account the real physiological respiratory dynamics. However, in [7] the authors present an ad-hoc evolutionary algorithm designed to explore a search space with 15 dimensions for the respiratory system including different organs. The method tries to estimate the parameters of a complex organ behavior model (15 parameters). The authors in [8] have proposed a FE model of the lung motion using a generic pressure-volume curve, which is not patient specific. Recently, the authors in [9] have proposed patient specific biomechanical model of the lung motion from 4D CT images for half respiratory cycle, where the motion is not constrained by any fixed boundary condition. The authors have used 4 and 16 pressure zones on the sub-diaphragm and thoracic cavity, respectively. Unfortunately, none of these methods take into account the real physiological respiratory properties, and are not able (or difficult) to be controlled or monitored by the external parameters. In this chapter, we evaluate the 3D tumor trajectories from patient-specific biomechanical models of the respiratory system for a whole respiratory cycle, based on personalized physiological pressure-volume curve [10]. This model has coupled an automatic tuning algorithm to calculate the personalized lung pressure and diaphragm force parameters.

2   Materials and Methods

2.1   Anatomy and Physiology of the Respiratory System

The lung is a passive organ which is divided into two halves, the right and left lung. It is situated in the thorax on either side of the heart. The pleural cavity is surrounded by the chest wall on the sides, and the diaphragm on the bottom. This space contains pleural fluid which facilitates near frictionless sliding at this boundary. The diaphragm is a dome-shaped musculofibrous membrane concave toward the lungs which separates the thorax from the abdominal cavity (Fig. 1). It is composed of a peripheral part (muscular fibre) and a central part (tendon). Lungs are linked to the diaphragm and to the ribs through the pleura. The mechanics of human breathing involves two steps that alternate with each other: inhalation (inspiration) and exhalation (expiration). Negative pressure in the pleural cavity (natural breathing) initiates when the diaphragm and chest wall move away from the lung. The negative pressure expands lung volume, dropping the internal lung pressure, allowing air to enter passively in the lung. The ability of the lungs to expand is expressed by using a measure known as the lung compliance. Lung compliance is the relationship between how much pressure is required to produce a degree of volume change of the lungs. It is affected by the elastic properties of the lung. The pulmonary compliance therefore reflects the lungs ability to develop in response to an increase in pressure.

2.2   3D Segmentation and CAD Reconstruction

Biomechanical modeling of the respiratory system necessitates the geometrical modeling of involved organs. For this purpose a correct segmentation of organs on CT images is necessary. Various approaches for multi-organ and lung segmentation have been developed based on CT images, which include gray-level thresholding,

Fig. 1 Respiratory mechanics: the role of the diaphragm and thorax in breathing

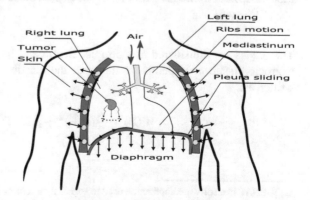

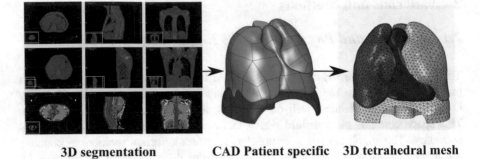

3D segmentation　　　　**CAD Patient specific**　**3D tetrahedral mesh**

Fig. 2 3D Segmentation, CAD reconstruction and 3D mesh patient specific adapted for finite element simulation

region growing, edge tracking. In this paper, the thorax, the lungs and the external skin are segmented automatically using gray-level thresholds algorithms available within ITK-SNAP library.[1] Automatic segmentation of the diaphragm is difficult due the lack of image contrast of the diaphragm with its surrounding organs as well as the respiration-induced motion artifacts in 4D CT images. The diaphragms were manually segmented within ITK-SNAP [11, 12]. In order to extract the mediastinum structure, we have used the different segmentation masks of the lungs, thorax, the inner thoracic region and the diaphragm. The accurate segmentation of lung tumors remains quite challenging, and the correct segmentation can only be achieved by medical experts.

After segmentation, a 3D surface mesh and a CAD-based approach has been developed. The organs shape are reconstructed as a solid using non-uniform rational B-spline (NURBS) curves. Using the resulting smooth surface, a quality mesh using a first-order tetrahedra elements (C3D4) is generated using Abaqus packages (Fig. 2).

2.3　Biomechanical Patient-Specific Model of the Respiratory System

The organs are considered as isotropic, elastic and hyperelastic materials. For an isotropic elastic or hyperelastic material, the elastic energy, denoted W, may be written as:

$$W(\mathbf{E}) = \frac{\lambda}{2}(tr\,\mathbf{E})^2 + \mu\,(tr\,\mathbf{E}^2) \tag{1}$$

---

[1]ITK-SNAP is a software application used to segment structures in 3D medical images.

where $\mathbf{E}$ is the Green-Lagrange strain tensor, λ and μ are the Lame coefficients. The Lame coefficients can be written in terms of Young's modulus, E, and Poisson's ratio, ν.

$$\mu = \frac{E}{2(1+\nu)} \qquad \lambda = \nu \frac{E}{(1-2\nu)(1+\nu)} \tag{2}$$

The second Piola-Kirchhoff stress tensor and the Green-Lagrange strain tensor given by:

$$\mathbf{S} = \lambda \, (tr \, \mathbf{E}) \, \mathbf{I} + 2 \mu \, \mathbf{E} \tag{3}$$

For dynamic simulation using FEM, the equation of motion of a vertex l of the organ mesh can be written:

$$M^l \{\ddot{\mathbf{u}}_l\} + \gamma^l \{\dot{\mathbf{u}}_l\} + \sum_{\tau \in \nu_l} \left(\left\{ \mathbf{F}_l^{int} \right\} \right) = \left\{ \mathbf{F}_{ext}^l \right\} \tag{4}$$

Where M^l, γ^l are respectively the mass and damping coefficients of each vertex. The ν_l is the neighborhood of vertex l (i.e., the tetrahedra containing node l). To solve the dynamic system, we have chosen the implicit finite difference scheme in time for more stability.

In our simulation, the mass density of each tissue is patient-specific, calculated and determined directly from CT scan images, based on the density mapping algorithm defined and developed in our previous works [14]: First, organs tetrahedral meshes are generated from segmented CT scanner images. Next, the Hounsfield values issued from CT scanner images are converted into density values that are mapped to the node of the mesh, respecting the principles of mass conservation (Fig. 3). For more information related to density mapping algorithm, one may refer to [13, 14].

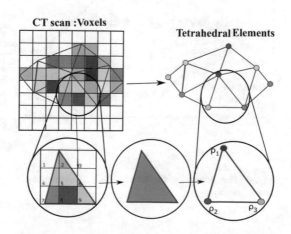

Fig. 3 Tetrahedral density map generation. The mass of a tetrahedral element equals the sum of the masses of volumes of intersection between the tetrahedron and the grid of voxels:
$m(T_k) = m(I_k^1) + m(I_k^2) + m(I_k^4) + m(I_k^5) + m(I_k^6) + m(I_k^8) + m(I_k^9)$

2.4 The Boundary Conditions

The developed biomechanical respiratory model is monitored directly by simulated actions of the breathing muscles; the diaphragm and the intercostal muscles/the rib cage. For the diaphragm, we have applied the radial direction of muscle forces, which corresponds anatomically to the direction of muscle fibers. The pressure is applied on the muscular part of the diaphragm and a simple homogeneous Dirichlet boundary conditions is applied in the lower part of the diaphragm and the Lagrange multiplier's method used for the contact model. In order to simulate the sliding of the lungs, a surface-to-surface contact model is applied on the lung-chest cavity, as well as lung-diaphragm cavity. The frictionless contact surfaces are used to simulate the pleural fluid behavior.

In our previous works [11, 12, 15], we have presented a methodology to study rib kinematics, using the finite helical axis method, where ribs could be considered as rigid bodies compared to other surrounding anatomical elements. The idea is to predict, from the transformation parameters, the rib positions and orientation at any time. Each rib transformation parameter is automatically computed between the initial and final states (Fig. 4). Then, we have applied a linear interpolation of the transformation to predict the rib motion at any intermediate breathing states. For more details about finite helical axis method, one can refer to [15].

In this work, the amplitude of the lung pressure and diaphragm force are patient specific, they are determined at different respiratory states by an optimization framework based on inverse finite element method [10]. The model is controlled by personalized pressure-volume curves (semi-static compliance), calculated by $C_{ss} = \frac{3(1-2\nu)}{E\,V_{t-1}}$ at different states. Where E, ν and V_{t-1} are Youngs modulus, Poisson coefficient and lung volume at step $t-1$ respectively. The mechanical properties and behaviors of the different organs used in our simulations are settled in the Table 1.

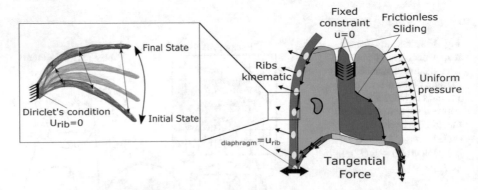

Fig. 4 The boundary conditions (BC) of our patient specific biomechanical model of the respiratory system

Table 1 Mechanical properties of breathing system: LE Linear Elastic, HVSK Hyperelastic Saint Venant Kirchhoff, E Youngs modulus, v Poisson coefficient, ρ volumetric density [10–12]

Tissues	Mechanical behavior	E (MPa)	v	ρ (kg/m³)
Lungs	HSVK	$3.74 * 10^{-3}$	0.3	$3 * 10^2$
Lung tumor	LE	49	0.4	$1.5 * 10^3$
Mediastinum	LE	$5.87 * 10^{-3}$	0.4	$1 * 10^2$
Diaphragm muscle	HSVK	5.32	0.33	$1 * 10^3$
Diaphragm tendon	LE	33	0.33	$1 * 10^3$
Ribs	LE	5000	0.3	$1.5 * 10^3$
Cartilage	LE	49	0.3	$1 * 10^3$
Body of sternum	LE	11,500	0.3	$1.5 * 10^3$
Thoracic vertebra	LE	9860	0.3	$1.5 * 10^3$
Flesh	LE	5.32	0.4	$1 * 10^6$

3   Results and Experimental Validation

3.1   Mesh Quality

The quality of the mesh plays a significant role in the accuracy and stability of the numerical computation. In our simulation, we have used the linear tetrahedral continuum elements (C3D4). These elements permit mesh refinement around areas of high stress concentration. By default, poor quality elements are those that fulfill one or several of the following criteria: jacobian greater than 0.6, ratio of the maximum side length to the minimum side length larger than 10, the shape factor ranges from 0 to 1, minimum interior angle smaller than 20 degrees, and maximum interior angle larger than 120 degrees.

In this chapter, the mesh quality Fig. 5 is performed using Abaqus packages. The above criteria for these elements are: 97, 83% of the elements with shape factor $\left(\frac{EV}{OEV}\right)^2$ between 0.1 and 1, 82, 95% elements with minimum angle $\geq$ 20, 99, 5% with maximum angle $\leq$ 140, 95, 9% with minimum length edge $\geq$ 3 mm, 99, 1% with maximum length edge $\leq$ 15 mm. From DIR-Lab Dataset [16], we have evaluated the motion estimation accuracy on two selected patients, with small and large breathing amplitudes (Patient 1 = 10.9 mm, Patient 10 = 26.06 mm). In our finite element simulation, we simulate the full breathing cycle, including 10 intermediate states (see Fig. 6). We define the simulation time for the inspiration phase is 2 s and for the expiration phase is 3 s. The Fig. 7 shows the displacement field of the lungs and diaphragm during breathing. For the diaphragm, we can observe the maximum displacement on the right-posterior (RP) and left-posterior (LP) sides. It is also possible to notice a slightly larger (RP) side motion than (LP)

---

[2]EV: element volume and OEV: Optimal element volume is the volume of an equilateral tetrahedron with the same circumradius as the element. (The circumradius is the radius of the sphere passing through the four vertices of the tetrahedron.)

Table 2 Average landmark lung error (mm) during exhalation at different respiratory states: the first state T00, the end inspiration (T50), the end expiration (T10)

Patients	Mean ± SD (mm)					Mean	Amplitude
	T10	T20	T30	T40	T50	All states	
Patient 1	2,0 ± 1,5	2,1 ± 1,2	2,1 ± 1,5	1,6 ± 1,3	1,2 ± 0,8	1,7 ± 1,3	10.9 (mm)
Patient 10	2,1 ± 1,5	2,2 ± 1,2	2,1 ± 1,6	1.6 ± 1,5	1.1 ± 0.8	1.8 ± 1.3	26,06 (mm)

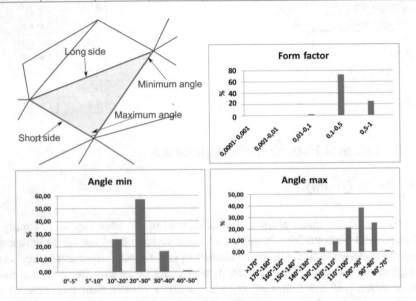

Fig. 5 Some criteria of mesh quality of tetrahedral elements. The triangular mesh element showing the longest side, shortest side, maximum interior angle and the minimum interior angle

side motion, according to the physiological anatomy. For the lungs deformation, the maximum displacement occurring in the posterior region along the superior-inferior (SI) direction (diaphragm direction).

Preliminary study was conducted to verify the efficiency of the developed finite element model and to evaluate lung tumor motion during full breathing cycle. In this order, the 3D lung tumor trajectories identified from 4D CT scan images were used as reference and compared with the 3D lung tumor trajectories estimated from finite element simulation during the whole cycle of breathing (10 phases between the EI and EE). The accuracy of the proposed tumor tracking method is evaluated by comparing and calculating the average Euclidean distance between the 3D mesh surface of the segmented tumor and predicted FE lung tumor. The Fig. 8 shows a comparison study between the hysteresis trajectories of the lung tumor during the whole cycle of the breathing compared to the trajectory calculated directly from 4D CT images. The results illustrate that our patient specific biomechanical model for tumor lung tracking is accurate and the average mean error is less than 1.8 ± 1.3 mm.

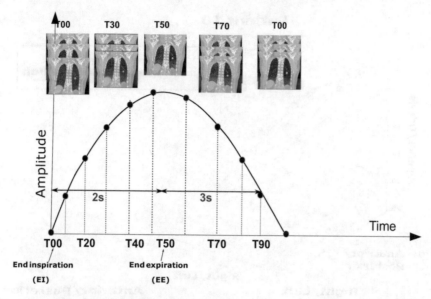

Fig. 6 Lung deformations during the full breathing cycle and intermediate states (10 states). Image slices of a patient case are taken from the DIR-lab data base [16]. The curve is only for illustration purposes

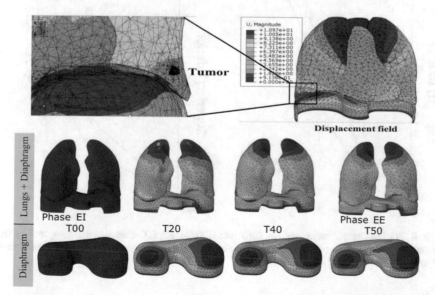

Fig. 7 Qualitative analysis of patient specific biomechanical simulation; lungs and diaphragm deformations from the end inhalation (EI) to end exhalation (EE), T00, T20, T40 and T50 are the intermediate states of the respiration between the EI to EE

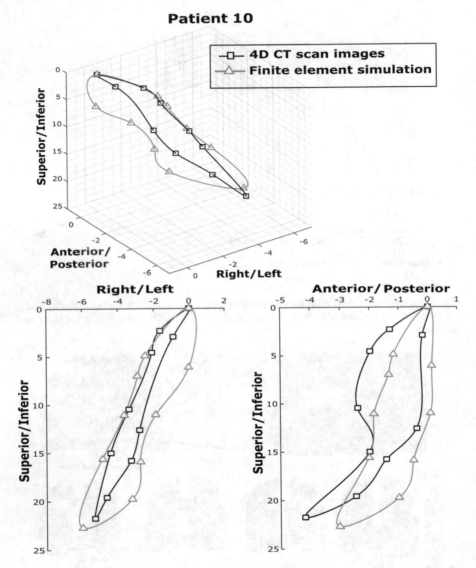

Fig. 8 3D lung tumor trajectory (in mm) issued from 4D CT scan images compared to the trajectory calculated by biomechanical finite element model including rib kinematics for patient P10 for DirLab data set [16]

4   Discussion and Conclusion

In this research work, we have developed a patient specific biomechanical model of the respiratory system for lung tumor tracking for the whole respiratory cycle. Our preliminary results are quite realistic compared to the 4D CT scan images. This could be a potential tool to provide valuable tumor motion information for physician to reduce the margins between clinical target volume (CTV) and planning target volume (PTV). One of the limitations of our work that the multiple organ shape reconstruction is time consuming and manual operations for each patient. In order to avoid manual contouring and 3D geometry segmentation for different organs, and to reduce the computational costs without lowering the quality, we plan to develop and use a realistic atlas-based 3D shape reconstruction of the respiratory system based on statistical training or machine learning, to get a fast and automatic patient-specific model. Also, the use of few patients is another limitation of the presented work. Future work could investigate more patients from DirLab data set [16] or other data bases. Currently, we are working on the optimization of our model. The goal is to produce a novel 4D computational patient specific model using non-invasive surrogates to predict and to monitor lung tumor motion during the treatment.

Acknowledgements This research is supported by the LABEX PRIMES (ANR-11-LABX-0063), within the program Investissements dAvenir(ANR-11-IDEX- 0007) operated by the French National Research Agency (ANR) and by France Hadron.

References

1. ICRU 50 Prescribing, Recording, and Reporting Photon Beam Therapy (ICRU Report, Technical Report vol. 50) (International Commission on Radiation Units and Measurements, Bethesda, 1993)
2. H. Shirato et al., Speed and amplitude of lung tumor motion precisely detected in four-dimensional setup and in real-time tumor-tracking radiotherapy. Int. J. Radiat. Oncol. Biol. Phys. **64**(4), 1229–1236 (2006)
3. J. Ehrhardt, C. Lorenz, *4D Modeling and Estimation of Respiratory Motion for Radiation Therapy* (Springer, Berlin/Heidelberg, 2013). ISBN: 978-3-642-36441-9
4. M. Behr, J. Pérès, M. Llari, Y. Godio, Y. Jammes, C. Brunet, A three-dimensional human trunk model for the analysis of respiratory mechanics. J. Biomech. Eng. **132**, 014501-1–014501-4 (2010)
5. M. Pato et al., Finite element studies of the mechanical behaviour of the diaphragm in normal and pathological cases. Comput. Methods Biomech. Biomed. Eng. **14**(6), 505–513 (2011)
6. A. Al-Mayah, J. Moseley, M. Velec, K. Brock, Toward efficient biomechanical-based deformable image registration of lungs for imageguided radiotherapy. Phys. Med. Biol. **56**(15), 4701 (2011)
7. F. Vidal, P.-F. Villard, E. Lutton, Tuning of patient specific deformable models using an adaptive evolutionary optimization strategy. IEEE Trans. Biomed. Eng. **59**(10), 2942–2949 (2012)
8. J. Eom et al., Predictive modelling of lung motion over the entire respiratory cycle using measured pressure-volume data, 4DCT images, and finite-element analysis. Med. Phys. **37**(8), 4389–4400 (2010)
9. B. Fuerst, T. Mansi, F. Carnis, M. Saelzle, T. Zhang, J. Declerck, T. Boettger, J. Bayouth, N. Navab, A. Kamen, Patient-specific biomechanical model for the prediction of lung motion from 4D CT images. IEEE Trans. Med. Imaging **34**(2), 599–607 (2015)

10. M. Giroux, H. Ladjal, M. Beuve, B. Shariat, in *Biomechanical Patient-Specific Model of the Respiratory System Based on 4D CT Scans and Controlled by Personalized Physiological.* Medical Image Computing and Computer Assisted Intervention – MICCAI 2017 – 20th International Conference, Quebec (Canada), 13 Sept 2017, pp. 216–223
11. H. Ladjal, B. Shariat, J. Azencot, M. Beuve, in *Appropriate Biomechanics and Kinematics Modeling of the Respiratory System: Human Diaphragm and Thorax* (IEEE, IROS, 2013)
12. H. Ladjal, J. Azencot, M. Beuve, P. Giraud, J.M. Moreau, B. Shariat, Biomechanical Modeling of the respiratory system: Human diaphragm and thorax In: Doyle B., Miller K., Wittek A., Nielsen P. (eds) Computational Biomechanics for Medicine. Springer, Cham, (2015) pp. 101–115 (15 p.) https://doi.org/10.1007/978-3-319-15503-6_10
13. P. Manescu, H. Ladjal, J. Azencot, M. Beuve, B. Shariat, Human liver multiphysics modeling for 4D dosimetry during hadrontherapy, in *IEEE 10th International Symposium on Biomedical Imaging (ISBI)*, 2013, pp. 472–475
14. P. Manescu, H. Ladjal, J. Azencot, M. Beuve, E. Testa, B. Shariat, Four-dimensional radiotherapeutic dose calculation using biomechanical respiratory motion description. Int. J. Comput. Assist. Radiol. Surg. **9**, 449–457 (2014)
15. A.L. Didier, P.F. Villard, J.Saade, J.M. Moreau, M. Beuve, B. Shariat, A chest wall model based on rib kinematics, in *IEEE ICV*, 2009, pp. 159–164
16. E. Castillo et al., Four-dimensional deformable image registration using trajectory modeling. Phys. Med. Biol. **55**, 305–327 (2009)

Design of Auxetic Coronary Stents by Topology Optimization

Huipeng Xue and Zhen Luo

Abstract Coronary artery stents are the most important implantation devices for the practice of the interventional cardiology to treat coronary artery disease (CAD) since the mid-1980s. However, the problems of stent thrombosis (ST) and in-stent restenosis (ISR) still exist. In addition to the reasons of implanted materials and coatings, mechanical and structural factors are also important factors and responsible for the complications, such as inadequate stent expansion, incomplete stent apposition and stent fracture in design. This research aims to develop a concurrent topology optimization by a parametric level set method associated with numerical homogenization method, to generate novel architectures for self-expanding (SE) stents with mechanical auxetic metamaterials. The topological design is firstly implemented in MATLAB, and then the optimized architecture is further improved and optimized in the commercial software ANSYS. The final stenting structure is numerically validated to demonstrate the effectiveness of the design method.

Keywords Self-expanding stents · Auxetics · Level sets · Topology optimization

1 Introduction

Coronary artery disease (CAD) also known as ischemic heart disease (IHD) has a high mortality even nowadays. Percutaneous coronary intervention (PCI) technology has been widely accepted as an effective treatment after 40 years development [1, 2]. Among that, the implantation of coronary stents can significantly decrease the rates of restenosis and abrupt closure of arteries to increase life expectancy of patients [3, 4].

H. Xue · Z. Luo (✉)
School of Mechanical and Mechatronic Engineering, University of Technology Sydney, Ultimo, NSW, Australia
e-mail: zhen.luo@uts.edu.au

© Springer Nature Switzerland AG 2020
K. Miller et al. (eds.), *Computational Biomechanics for Medicine*,
https://doi.org/10.1007/978-3-030-42428-2_3

17

In the early days, bare-metal stents (BMS) were used in conjunction with angioplasty due to successful results in treating abrupt and susceptible vessel closure [5, 6]. However, the incidence of stent thrombosis (ST), in-stent restenosis (ISR) and other complications [7] resulted in the generation of drug-eluting stents (DES) [8]. DES are superior to BMS in that it can reduce the rate of ISR but have a higher risk of ST in the late thrombosis [9, 10], due to drug coatings. Even for the new generation of biodegradable stents (BDS) and bioresorbable vascular scaffolds (BVS), these drawbacks still remain [11, 12]. Compared with the risk of ST in the late healing stage, DES show an obvious decrease of ISR in the short-term treatment without brachytherapy or intracoronary radiation. This is the reason why DES are more popular recently. Nevertheless, it has been reported that DES result in a higher risk of late thrombosis compared with BMS. The much higher cost of DES doesn't lead to a significant increase in life expectancy than other stents [13].

According to different expansion mechanisms, stents can also be divided into self-expanding (SE) and balloon-expandable (BE) stents. In 1986, stents with self-expanding properties were firstly introduced into balloon angioplasty for treating abrupt closure of arteries [3]. The characteristics of positive supporting and shape memory metal materials [14] gave good short-term treatment results. The most advantages of SE stents can be summarized as: (1) The gradual expansion manner of SE stents leads to a lower incidence of edge dissections. It can avoid immediate vessel wall injury compared with BE stents, which makes SE stents more suitable for treating small-diameter vessels [15], (2) The good conformability makes it easily to match different lesion shapes, which is superior to any other stent for treating vulnerable plaques and bifurcation lesions, as well as preventing inadequate stent expansion, and (3) The used superelastic materials exhibit much better mechanical properties than materials of BE stents with respect to fracture toughness, flexibility, fatigue strength and corrosion resistance.

However, some unfavorable features [16, 17] of SE stents limit their clinical use. First, the SE stents are usually hosed into cumbersome catheters during the implantation, which makes the delivery difficult. Second, the complicated placement demands high accuracy due to the phenomenon of foreshortening after deployment. Third, the continual outward supporting of conventional SE stents is not adaptive and difficult to accurately control, which may lead to a larger luminal diameter than the original size that will further pose a thrombotic threat.

Besides biological factors, structural or mechanical aspects also play an important role in stents, and they can trigger serious complications finally leading to ST and ISR, such as inadequate stent expansion, incomplete stent apposition and stent fracture in design [18]. These issues can be addressed via new stenting structures, new artificial materials or new expansion methods. Hence, the alternative designs that can avoid or help reduce these complications are still in demands.

In this paper, we will focus on the development of a novel family of SE stents using topological design optimization technology together with a new type of mechanical metamaterials-auxetics, with a view to generating new stenting structural architectures, to help reduce the occurrence of ST and ISR after implantation.

Compared to most conventional materials with positive Poisson's ratios, auxetics are a special kind of mechanical metamaterials artificially designed to exhibit negative Poisson's ratios (NPR) [19, 20]. Auxetic materials will contract in transverse directions when they are compressed uniaxially. Auxetics provides enhanced mechanical properties such as indentation resistance, fracture toughness, and shear stiffness, which greatly facilitate a range of applications, including energy absorption, anti-impact, thermal isolation and biomedical applications [21, 22].

Topology optimization provides an efficient way to find the best material distributions under the boundary and loads conditions in the design domain. It has been wildly used in the structural and material designs over the past two decades, and several popular methods have been developed, such as the solid isotropic material with penalization (SIMP) method [23, 24], the evolutionary structural optimization (ESO) method [25], and level set method (LSM) [26–28].

The numerical homogenization method [29, 30] has been developed to evaluate the effective properties of microstructures. It is usually combined with other topology optimization methods for the design of microstructures and the related cellular composites. This kind of cellular composites mostly consists of periodic microstructures and the microstructures can be given special properties such as auxetics. The topological design of multifunctional cellular composites enables many applications in engineering [31].

LSM is one of the recently developed method for topological shape optimization of structures. It has shown excellent ability to capture geometry and shape of the design. The key concept is to embed the design boundary of a structure as the zero-level set of a higher-dimensional level set function. Since the evolution of the level set function can be described by the Hamilton-Jacobi Partial Differential Equation (PDE) [32], the dynamic motion of the level set function can be tracked by solving this equation. However, some strict conditions are required during the numerical implementation of the H-J PDE, such as the Courant-Friedrichs-Lewy (CFL) condition, boundary velocity extensions, and re-initializations [32]. As one of the alternative LSMs, the parametric level set method (PLSM) [33–35] has shown it is high efficiency in solving topology optimization [36] and this paper will apply the PLSM to design the stenting structural architectures.

To realize the design of ASE stents, a concurrent topological design method will be applied to find auxetic stenting architecture as microstructures, and at the same time the compliance of the macro stenting structure is considered to maintain the stiffness requirement of stents. Topology optimization will be applied to explore the best material layout for the SE stents, and the auxetics will be included into the biocompatible materials to enable an adaptive "self-expanding" procedure of stenting structures. The structure periodically consists of identical auxetic unit cells. This will deliver a new kind of auxetic SE (ASE) stents to address the above problems relevant to ST and ISR due to the mechanical and structural issues of the current stenting designs. The topological optimization can help find the most efficient stenting structures, and auxetics will make SE stents much smaller when compressed, beneficial to deliverability. The optimized ASE stents can also

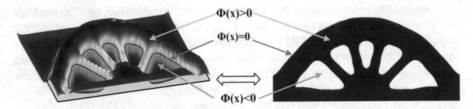

Fig. 1 Level set function (left) and design domain located at zero level set (right)

eliminate the foreshortening to help the deployment. Moreover, the auxetic behavior can also enhance the flexibility, conformability, and fatigue strength of SE stents.

2   Parametric Level-Set Method

The unique characteristic of the level set method is the implicit description of the structural boundary which is presented at the zero level set of a higher dimensional level set function $\Phi(x)$, as shown in Eq. (1) As a 2D example illustrated in Fig. 1, $\Phi(x) = 0$ shows the boundary of a structure located at zero level set.

$$\begin{cases} \Phi(x) > 0 & x \in \Omega \backslash \partial\Omega & (Material) \\ \Phi(x) = 0 & x \in \partial\Omega & (Boundary) \\ \Phi(x) < 0 & x \in D \backslash (\Omega \cup \partial\Omega) & (Void) \end{cases} \tag{1}$$

where x is the point in the space D, Ω and $\partial\Omega$ denote the design domain and the boundary, respectively. The dynamic motion of the design domain Ω can be achieved by solving Hamilton-Jacobi PDE, as shown in (2). In that process, the normal velocity filed V_n of the boundary $\partial\Omega$ is used to enable the dynamic motion of the level set function.

$$\frac{\partial \Phi(x,t)}{\partial t} - V_n |\nabla\Phi(x,t)| = 0 \tag{2}$$

The interpolation of the level set function $\Phi(x)$ by using CSRBFs $\varphi(x)$ based on the fixed knots in the design domain can be described as Eq. (3).

$$\Phi(x,t) = \varphi(x)^T \alpha(t) = \sum_{i=1}^{N} \varphi_i(x)\alpha_i(t) \tag{3}$$

where N is the total number of fixed knots in the design domain, $\alpha_i(t)$ is the expansion coefficient of the interpolation with respect of the ith knot, and the CSRBFs of the ith knot used with C2 continuity is given by:

$$\varphi_i(x) = \max\left\{0, (1 - r_i(x))^4\right\}(4r_i(x) + 1)$$

$$r_i(x) = d_I/d_{mI} = \sqrt{(x - x_i)^2 + (y - y_i)^2}/d_{mI} \tag{4}$$

where d_I denotes the distance between the current sample knot (x, y) and the ith knot (x_i, y_i), and d_{mI} denotes the radius of the support domain of the ith knot.

Then, the conventional Hamilton–Jacobi PDE is transformed as Eq. (5), and the new velocity field V_n can be described as (6). Therefore, the dynamic motion of level set function $\Phi(x)$ is only related to the design variables expansion coefficient vector $\alpha(t)$. Because $\alpha(t)$ is being evaluated by all knots in the design domain, no addition extension scheme is required. In this way, the standard LSM is converted into a parametric form.

$$\varphi(X)^T \dot{\alpha}(t) - V_n \left|(\nabla\varphi)^T \alpha(t)\right| = 0 \tag{5}$$

$$V_n = \frac{\varphi(X)^T}{\left|(\nabla\varphi)^T \alpha(t)\right|}\dot{\alpha}(t), \quad \text{where } \dot{\alpha}(t) = \frac{d\alpha(t)}{dt} \tag{6}$$

3 Numerical Homogenization Method

The numerical homogenization method has been widely used to approximate the effective properties of microstructures. The effective elasticity tensor $DH\ ijkl$ of a 2D microstructure can be calculated by:

$$D_{ijkl}^H = \frac{1}{|\Omega^{MI}|} \int_{\Omega^{MI}} \left(\varepsilon_{pq}^{0(ij)} - \varepsilon_{pq}^{*(ij)}\left(u^{MI(ij)}\right)\right)$$

$$\times D_{pqrs}\left(\varepsilon_{rs}^{0(kl)} - \varepsilon_{rs}^{*(kl)}\left(u^{MI(ij)}\right)\right) H\left(\Phi^{MI}\right) d\Omega^{MI} \tag{7}$$

where the superscript 'MI' indicates the quantities in the microscale; Ω^{MI} is the design domain of the microstructure; $|\Omega^{MI}|$ is the area of the microstructure; and Φ^{MI} is the level set function in the microscale. $i, j, k, l = 1, 2$. D_{pqrs} is the elasticity tensor of the base material. $H(\Phi^{MI})$ is the Heaviside function [27]. $\varepsilon_{pq}^{0(ij)}$ is the test unit strain field, where $(1,0,0)^T$, $(0,1,0)^T$ and $(0,0,1)^T$ are used in 2D models; $\varepsilon_{pq}^{*(ij)}$ is the locally varying strain fields and defined by:

$$\varepsilon_{pq}^{*(ij)}\left(u^{MI(ij)}\right) = \frac{1}{2}\left(u_{p,q}^{MI(ij)} + u_{q,p}^{MI(ij)}\right) \tag{8}$$

By using the virtual displacement field $v^{MI(kl)}$ in $\overline{U}\left(\Omega^{MI}\right)$ that is the space consisting of all the kinematically admissible displacements in Ω^{MI}, the displacement field $u^{MI(ij)}$ can be calculated through finite element analysis using the periodical boundary conditions of the microstructure:

$$\int_{\Omega^{MI}}\left(\varepsilon_{pq}^{0(ij)}-\varepsilon_{pq}^{*(ij)}\left(u^{MI(ij)}\right)\right)D_{pqrs}\varepsilon_{rs}^{*(kl)}\left(v^{MI(kl)}\right)$$
$$\times H\left(\Phi^{MI}\right)d\Omega^{MI}=0, \quad \forall \ v^{MI(kl)} \in \overline{U}\left(\Omega^{MI}\right)$$

(9)

4 The First Optimization Stage for the Design of Auxetics

4.1 The Concurrent Optimization Scheme

The concurrent topology optimization scheme is defined as a multi-objective optimization problem to find an expansion coefficient vector α_n^{MI} for microstructure to obtain negative Poisson's ratios, and minimum the compliance of the macrostructure. A piece of the stent approximated as rectangle shape is used as the micro design domain consisted of one unique microstructure, shown in Fig. 2; two coordinates are used to describe the design domains: the macrostructure(X_1, X_2) and microstructure(Y_1, Y_2); the vertical degree of freedom is fixed at the top and bottom edges of the macro structure, while two unit forces F are applied on the left and right edges in the horizontal direction. 2D four-node rectangle elements is adopted and each element has a unit length, height. The artificial base material model with Young's modulus 1 and Poisson's ratio 0.3 used. The numerical design scheme can be described as Eq. (10).

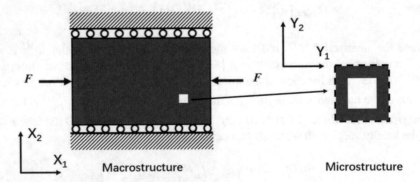

Fig. 2 The macrostructure(left) and microstructure(right)

Find $\quad \alpha_n^{MI} \ (n = 1, 2, \ldots, N)$

Min $\quad J = J^{MA} + J^{MI}$

S.T. $\quad G = \int_{\Omega^{MI}} H\left(\Phi^{MI}\right) d\Omega^{MI} \leq V^{max}$

$\quad\quad a^{MA}\left(u^{MA}, v^{MA}\right) = l^{MA}\left(v^{MA}\right), \forall v^{MA} \in \overline{U}\left(\Omega^{MA}\right)$

$\quad\quad a^{MI}\left(u^{MI}, v^{MI}, \Phi^{MI}\right) = l^{MA}\left(v^{MA}, \Phi^{MI}\right), \forall v^{MI} \in \overline{U}\left(\Omega^{MI}\right)$ $\quad\quad$ (10)

$\quad\quad \alpha_{min}^{MI} \leq \alpha_n^{MI} \leq \alpha_{max}^{MI}$

where,

$$J^{MA} = \tfrac{1}{2}\int_{\Omega^{MA}} \varepsilon_{ij}\left(u^{MA}\right) D_{ijkl}^H \varepsilon_{kl}\left(u^{MA}\right) d\Omega^{MA}$$
$$J^{MI} = \left(D_{12}^H/D_{11}^H + 1\right)^2 + \left(D_{12}^H/D_{22}^H + 1\right)^2$$

where, the superscript 'MA' and 'MI' denotes the macro and micro quantities, respectively. The expansion coefficients of the CSRBF interpolation α_n^{MI} is the design variable in the microscale, which are within α_{min}^{MI} and α_{max}^{MI}. N is the total number of fixed knots in the micro design domain. J is the total objective function, which is comprised of the macro objective function J^{MA} the compliance of the macrostructure, and micro objective function J^{MI} the Poisson's ratios of the microstructure. D_{11}^H, D_{12}^H, D_{22}^H are specific values of the effective elasticity tensor of the microstructure. Here, the optimized microstructure is defined as isotropic or orthotropic material, thus there are two Poisson's ratios defined the in micro objective function. G is the volume constraint and the upper limitation is defined as V^{max}. u and v are the real and virtual displacement fields.

The bilinear energy and the linear load forms of the finite element model in the macroscale can be described as:

$$a^{MA}\left(u^{MA}, v^{MA}\right) = \int_{\Omega^{MA}} \varepsilon_{ij}\left(u^{MA}\right) D_{ijkl}^H \varepsilon_{kl}\left(v^{MA}\right) d\Omega^{MA} \quad\quad (11)$$

$$l^{MA}\left(v^{MA}\right) = \int_{\Omega^{MA}} p v^{MA} d\Omega^{MA} + \int_{\Omega^{MA}} \tau v^{MA} d\Gamma^{MA} \quad\quad (12)$$

where p is the body force and τ is the traction of the boundary Γ^{MA}. The bilinear energy and the linear load forms of finite element model in the microscale can be described as:

$$a^{MI}\left(u^{MI}, v^{MI}, \Phi^{MI}\right)$$
$$= \int_{\Omega^{MI}} \varepsilon_{ij}^{*(ij)}\left(u^{MI(ij)}\right) D_{pqrs} \varepsilon_{kl}^{*(kl)}\left(v^{MI(kl)}\right) H\left(\Phi^{MI}\right) d\Omega^{MI} \quad\quad (13)$$

$$l^{MI}\left(v^{MI}, \Phi^{MI}\right) = \int_{\Omega^{MI}} \varepsilon_{ij}^{0(ij)} D_{pqrs} \varepsilon_{kl}^{*(kl)}\left(v^{MI(kl)}\right) H\left(\Phi^{MI}\right) d\Omega^{MI} \quad\quad (14)$$

4.2   The Sensitivity Analysis

Based on the concurrent topology optimization model presented in Sect. 4.1, the sensitivity analysis of the design variables is required. It is divided into two parts due to the two different scales and calculated based on the first-order derivatives of the objective functions with respect to the expansion coefficients α_n^{MI}. The sensitivity in the macro-scale is:

$$\frac{\partial J^{MA}}{\partial \alpha_n^{MI}} = \frac{1}{2} \int_{\Omega^{MA}} \varepsilon_{ij}\left(u^{MA}\right) \frac{\partial D_{ijkl}^H}{\partial \alpha_n^{MI}} \varepsilon_{kl}\left(u^{MA}\right) d\Omega^{MA} \tag{15}$$

Since the elastic system is self-adjoint [37], the shape derivative of the elasticity tensor D_{ijkl}^H can be calculated by:

$$\frac{\partial D_{ijkl}^H}{\partial t} = -\frac{1}{|\Omega^{MI}|} \int_{\Omega^{MI}} \beta\left(u^{MI}\right) \varphi^{MI}(x)^T V_n \left|\left(\nabla \Phi^{MI}\right)^T\right| \delta\left(\Phi^{MI}\right) d\Omega^{MI} \tag{16}$$

where $\delta(\Phi^{MI})$ is the derivative of the Heaviside function $H(\Phi^{MI})$, and $\beta(u^{MI})$ is:

$$\beta\left(u^{MI}\right) = \left(\varepsilon_{pq}^{0(ij)} - \varepsilon_{pq}^{*(ij)}\left(u^{MI(ij)}\right)\right) D_{pqrs}\left(\varepsilon_{rs}^{0(kl)} - \varepsilon_{rs}^{*(kl)}\left(u^{MI(kl)}\right)\right) \tag{17}$$

Substituting the normal velocity V_n^{MI} defined in Eq. (6) into Eq. (17):

$$\frac{\partial D_{ijkl}^H}{\partial t} = -\sum_{n=1}^N \left(\frac{1}{|\Omega^{MI}|} \int_{\Omega^{MI}} \beta\left(u^{MI}\right) \varphi^{MI}(x)^T \delta\left(\Phi^{MI}\right) d\Omega^{MI}\right) \dot{\alpha}_n^{MI}(t) \tag{18}$$

While, the first-order derivative of the effective elasticity tensor D_{ijkl}^H with respect to t can be directly obtained by the chain rule:

$$\frac{\partial D_{ijkl}^H}{\partial t} = \sum_{n=1}^N \frac{\partial D_{ijkl}^H}{\partial \alpha_n^{MI}} \dot{\alpha}_n^{MI}(t) \tag{19}$$

Comparing (18) and (19), the derivative of the effective elasticity tensor D_{ijkl}^H with respect to the design variables α_n^{MI} can be calculated as:

$$\frac{\partial D_{ijkl}^H}{\partial \alpha_n^{MI}} = -\frac{1}{|\Omega^{MI}|} \int_{\Omega^{MI}} \beta\left(u^{MI}\right) \varphi^{MI}(x)^T \delta\left(\Phi^{MI}\right) d\Omega^{MI} \tag{20}$$

Then the derivative of the macro objective function J^{MA} with respect to the design variables α_n^{MI} can be obtained by Substituting Eq. (20) into (15). Similarly, the derivative of the micro objective function J^{MI} with respect to the design variables can be calculated, as shown in (21), and the derivative of the volume constrains G with respect to the design variables are given by (22).

$$\frac{\partial J^{MI}}{\partial \alpha_n^{MI}} = \frac{\partial \left(D_{12}^H / D_{11}^H + 1 \right)^2}{\partial \alpha_n^{MI}} + \frac{\partial \left(D_{12}^H / D_{22}^H + 1 \right)^2}{\partial \alpha_n^{MI}} \tag{21}$$

$$\frac{\partial G}{\partial \alpha_n^{MI}} = \int_{\Omega^{MI}} \varphi^{MI}(x)^T \delta \left(\Phi^{MI} \right) d\Omega^{MI} \tag{22}$$

4.3 Numerical Results

One of the main purposes of ISR is to implanting materials into the vessels, so the design of a stent usually uses as less material as possible to decrease the contacts between the stent and vessel walls. Meanwhile, the volume fraction of 35% is used for the microstructure design to ensure structural stiffness. To evaluate the numerical result, two values of Poisson's ratios *Mu1* and *Mu2* in two directions are defined as Eq. (23).

$$Mu1 = D_{12}^H / D_{11}^H, \quad Mu2 = D_{12}^H / D_{22}^H \tag{23}$$

Different discretized size of micro design domain will lead to different results, that is because more elements used in the design domain may capture more details of the optimized structure. Therefore, three different size of discretization 60×60, 100×100, 40×40 are used, and the relevant results are list in the Table 1. All three results are of clear and smooth boundaries of the microstructures, and exhibit NPR properties in both two directions. The material of the stent should be uniformly distributed. In the result of 100×100, the bridges in the middle, top and bottom are too thin compared with other parts, so this is not a very good choice.

Mu1 and *Mu2* are used to illustrate the Poisson's ratios in two directions, where *Mu1* can be used to evaluate the deformation along the horizontal direction when deformed in the vertical direction, and *Mu2* is used to describe the opposite situation. Although both negative values of *Mu1* and *Mu2* are desired to obtain a smaller volume of stent when compressed, a smaller absolute value of *Mu1* can lead to a smaller deformation in the axis direction when stent supporting the vessel, which will prevent the shortening of the stent in axis direction. Hence, the result of 40×40 is better than 60×60. From that, we can see more elements may capture more details of the structure, but it may also lead to a complex or ununiform distribution of material which may not suitable for the stent design.

Table 1 Three numerical optimization results for auxetic microstructures

Size	Microstructure	Level set surface	Effective elasticity tensor	Poisson's ratio
40 × 40			$\begin{bmatrix} 0.1473 & -0.0352 & 0 \\ -0.0352 & 0.0342 & 0 \\ 0 & 0 & 0.0043 \end{bmatrix}$	$Mu1 = -0.2390$ $Mu2 = -1.0292$
60 × 60			$\begin{bmatrix} 0.0578 & -0.0362 & 0 \\ -0.0362 & 0.0395 & 0 \\ 0 & 0 & 0.0023 \end{bmatrix}$	$Mu1 = -0.6263$ $Mu2 = -0.9165$
100 × 100			$\begin{bmatrix} 0.0509 & -0.0438 & 0 \\ -0.0438 & 0.0518 & 0 \\ 0 & 0 & 0.0021 \end{bmatrix}$	$Mu1 = -0.8605$ $Mu2 = -0.8456$

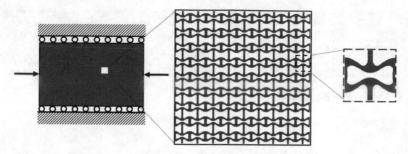

Fig. 3 The macro structure (left), 9 × 9 microstructures (middle), and the unit cell of microstructure (right)

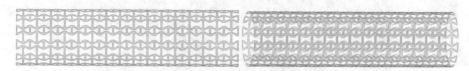

Fig. 4 The stent structure built by the first design result

The optimized structure of 40 × 40 element scale is adopted in the first numerical optimization stage, as shown in Fig. 3. From the figure, we can see the microscale is much smaller than the macroscale. However, as mentioned before, if much smaller microstructures are used, the one piece of the stent will be fully filled with the material as the left figure shown in Fig. 3. By doing this, the flexibility and conformability of the stent will decrease, and the incidence of ISR will significantly increase. Therefore, the optimized microstructure will be regarded as a smaller periodical macro unit cell in the macroscale.

5 The Second Optimization Stage and the Numerical Validation

Since the mechanical behavior of a stent is more similar to a shell that the dimension of the thickness is much smaller than the dimensions of the length and width. 2D four-node rectangle element is used in the first step due to the computational efficiency, while the shell element needs to be adopted in the second stage of the optimization to amend the accuracy of the final design. Thus, the commercial software ANSYS v19.2 is utilized to preform topology optimization for a stent again with shell elements, based on the optimized result from the first stage. The geometry is built by 12 unit cells along the circumference and 16 unit cells along the axis, and 10 times bigger than the real stent as shown in Fig. 4.

The volume fraction of the microstructure is specified as 35% in the first stage, and not too much material needs to be removed in the current stage. Hence, 10%

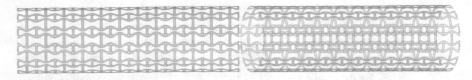

Fig. 5 The result of the second topology optimization

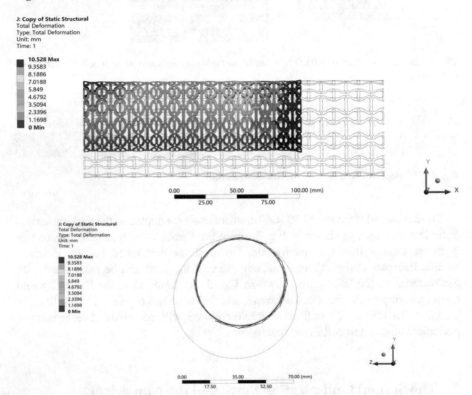

Fig. 6 The pression test: the front view(left) and the right view(right)

volume fraction is used to maximum the global compliance of the stent in the second stage. The optimized result can be seen in Fig. 5, and we can see small holes are generated in all the joints of the unit cells.

The numerical validation is performed to test the auxetic property of the optimized stent in ANSYS. In the simulation, the degree of freedom in the X direction of the left edge and one point in the left end is fixed and a force applied on the right edge of the stent to compress or stretch it. The test under pression is shown in Fig. 6. The colourful structure shows deformed stent, while the grey colour shows undeformed stent. From the figure we can see the stent contract in the radial directions when they are compressed uniaxially. In the right-side view, the diameter become smaller compared with the original size of the stent.

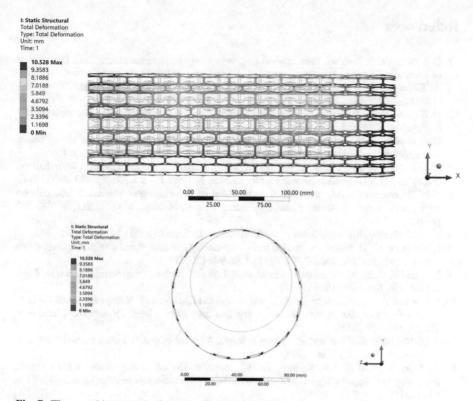

Fig. 7 The stretching test: the front view(left) and the right view(right)

Then, a stretching test is also performed, the result as shown in Fig. 7. The stent expanded in the radial directions when they are stretched uniaxially. Therefore, both compression and stretching test performed for the optimized stent illustrate a significant auxetic property.

6 Conclusion

The properties of auxetic structures can well satisfy the mechanic requirements of SE coronary artery stents and enhance their abilities of dealing with the mechanical factors of ST and ISR. The stent design using parametric level set topology optimization method provides a concurrent design of both material microstructures and macro meta-structure, which benefits the stent designs for applications in practice. However, another important characteristic of the materials of SE stents is the property of shape memory, and it will influence the deformation mechanism during the expanding. Therefore, the shape memory behaviour may need to be integrated into the auxetic design of SE stent in the near future.

References

1. C.T. Dotter, M.P. Judkins, Transluminal treatment of arteriosclerotic obstruction: Description of a new technic and a preliminary report of its application. Circulation **30**(5), 654–670 (1964)
2. A. Grüntzig, Transluminal dilatation of coronary-artery stenosis. Lancet **311**(8058), 263 (1978)
3. U. Sigwart et al., Intravascular stents to prevent occlusion and re-stenosis after transluminal angioplasty. N. Engl. J. Med. **316**(12), 701–706 (1987)
4. G.S. Roubin et al., Intracoronary stenting for acute and threatened closure complicating percutaneous transluminal coronary angioplasty. Circulation **85**(3), 916–927 (1992)
5. P.W. Serruys et al., A comparison of balloon-expandable-stent implantation with balloon angioplasty in patients with coronary artery disease. N. Engl. J. Med. **331**(8), 489–495 (1994)
6. D.L. Fischman et al., A randomized comparison of coronary-stent placement and balloon angioplasty in the treatment of coronary artery disease. N. Engl. J. Med. **331**(8), 496–501 (1994)
7. D. J. Moliterno, *Healing Achilles—sirolimus versus paclitaxel.* (2005), Mass Medical Soc.
8. C. Stettler et al., Outcomes associated with drug-eluting and bare-metal stents: A collaborative network meta-analysis. Lancet **370**(9591), 937–948 (2007)
9. L. Mauri et al., Stent thrombosis in randomized clinical trials of drug-eluting stents. N. Engl. J. Med. **356**(10), 1020–1029 (2007)
10. S. Cook et al., Correlation of intravascular ultrasound findings with histopathological analysis of thrombus aspirates in patients with very late drug-eluting stent thrombosis. Circulation **120**(5), 391–399 (2009)
11. S. McMahon et al., Bio-resorbable polymer stents: A review of material progress and prospects. Prog. Polym. Sci. **83**, 79–96 (2018)
12. A. Siiki, J. Sand, J. Laukkarinen, A systematic review of biodegradable biliary stents: Promising biocompatibility without stent removal. Eur. J. Gastroenterol. Hepatol. **30**(8), 813–818 (2018)
13. S.-H. Kang et al., Biodegradable-polymer drug-eluting stents vs. bare metal stents vs. durable-polymer drug-eluting stents: A systematic review and Bayesian approach network meta-analysis. Eur. Heart J. **35**(17), 1147–1158 (2014)
14. T. Duerig, D. Tolomeo, M. Wholey, An overview of superelastic stent design. Minim. Invasive Ther. Allied Technol. **9**(3–4), 235–246 (2000)
15. M. Joner et al., Histopathologic evaluation of nitinol self-expanding stents in an animal model of advanced atherosclerotic lesions. EuroIntervention **5**(6), 737–744 (2010)
16. R.A. Schatz et al., Clinical experience with the Palmaz-Schatz coronary stent. Initial results of a multicenter study. Circulation **83**(1), 148–161 (1991)
17. S. Garg, P.W. Serruys, Coronary stents: Looking forward. J. Am. Coll. Cardiol. **56**(10 Supplement), S43–S78 (2010)
18. S. Garg, P.W. Serruys, Coronary stents: Current status. J. Am. Coll. Cardiol. **56**(10 Supplement), S1–S42 (2010)
19. R. Lakes, Advances in negative Poisson's ratio materials. Adv. Mater. **5**(4), 293–296 (1993)
20. K.E. Evans, A. Alderson, Auxetic materials: Functional materials and structures from lateral thinking! Adv. Mater. **12**(9), 617–628 (2000)
21. R. Lakes, Response: Negative Poisson's ratio materials. Science **238**(4826), 551–551 (1987)
22. W. Yang et al., Review on auxetic materials. J. Mater. Sci. **39**(10), 3269–3279 (2004)
23. M. Zhou, G. Rozvany, The COC algorithm, part II: Topological, geometrical and generalized shape optimization. Comput. Methods Appl. Mech. Eng. **89**(1–3), 309–336 (1991)
24. M.P. Bendsøe, O. Sigmund, Material interpolation schemes in topology optimization. Arch. Appl. Mech. **69**(9–10), 635–654 (1999)
25. Y.M. Xie, G.P. Steven, A simple evolutionary procedure for structural optimization. Comput. Struct. **49**(5), 885–896 (1993)

26. J.A. Sethian, A. Wiegmann, Structural boundary design via level set and immersed interface methods. J. Comput. Phys. **163**(2), 489–528 (2000)
27. M.Y. Wang, X. Wang, D. Guo, A level set method for structural topology optimization. Comput. Methods Appl. Mech. Eng. **192**(1–2), 227–246 (2003)
28. G. Allaire, F. Jouve, A.-M. Toader, Structural optimization using sensitivity analysis and a level-set method. J. Comput. Phys. **194**(1), 363–393 (2004)
29. J. Guedes, N. Kikuchi, Preprocessing and postprocessing for materials based on the homogenization method with adaptive finite element methods. Comput. Methods Appl. Mech. Eng. **83**(2), 143–198 (1990)
30. O. Sigmund, Materials with prescribed constitutive parameters: An inverse homogenization problem. Int. J. Solids Struct. **31**(17), 2313–2329 (1994)
31. J. Chen, Y. Huang, M. Ortiz, Fracture analysis of cellular materials: A strain gradient model. J. Mech. Phys. Solids **46**(5), 789–828 (1998)
32. S. Osher, J.A. Sethian, Fronts propagating with curvature-dependent speed: Algorithms based on Hamilton-Jacobi formulations. J. Comput. Phys. **79**(1), 12–49 (1988)
33. Z. Luo et al., Shape and topology optimization of compliant mechanisms using a parameterization level set method. J. Comput. Phys. **227**(1), 680–705 (2007)
34. Z. Luo et al., A level set-based parameterization method for structural shape and topology optimization. Int. J. Numer. Methods Eng. **76**(1), 1–26 (2008)
35. Z. Luo, L. Tong, Z. Kang, A level set method for structural shape and topology optimization using radial basis functions. Comput. Struct. **87**(7–8), 425–434 (2009)
36. H. Wendland, Computational aspects of radial basis function approximation, in *Studies in Computational Mathematics*, (Elsevier, UK, 2006), pp. 231–256
37. X. Wang, Y. Mei, M.Y. Wang, Level-set method for design of multi-phase elastic and thermoelastic materials. Int. J. Mech. Mater. Des. **1**(3), 213–239 (2004)

Physics-Based Deep Neural Network for Real-Time Lesion Tracking in Ultrasound-Guided Breast Biopsy

Andrea Mendizabal, Eleonora Tagliabue, Jean-Nicolas Brunet, Diego Dall'Alba, Paolo Fiorini, and Stéphane Cotin

Abstract In the context of ultrasound (US) guided breast biopsy, image fusion techniques can be employed to track the position of US-invisible lesions previously identified on a pre-operative image. Such methods have to account for the large anatomical deformations resulting from probe pressure during US scanning within the real-time constraint. Although biomechanical models based on the finite element (FE) method represent the preferred approach to model breast behavior, they cannot achieve real-time performances. In this paper we propose to use deep neural networks to learn large deformations occurring in ultrasound-guided breast biopsy and then to provide accurate prediction of lesion displacement in real-time. We train a U-Net architecture on a relatively small amount of synthetic data generated in an offline phase from FE simulations of probe-induced deformations on the breast anatomy of interest. Overall, both training data generation and network training are performed in less than 5 h, which is clinically acceptable considering that the biopsy can be performed at most the day after the pre-operative scan. The method is tested both on synthetic and on real data acquired on a realistic breast phantom. Results show that our method correctly learns the deformable behavior modelled via FE simulations and is able to generalize to real data, achieving a target registration error comparable to that of FE models, while being about a hundred times faster.

Keywords Ultrasound-guided breast biopsy · Deep neural networks · Real-time simulation

A. Mendizabal and E. Tagliabue contributed equally to the paper.

A. Mendizabal (✉) · J.-N. Brunet · S. Cotin
INRIA Strasbourg, Strasbourg, France
e-mail: andrea.mendizabal@inria.fr; stephane.cotin@inria.fr

E. Tagliabue (✉) · D. Dall'Alba · P. Fiorini
Università degli Studi di Verona, Verona, Italy
e-mail: eleonora.tagliabue@univr.it

K. Miller et al. (eds.), *Computational Biomechanics for Medicine*,
https://doi.org/10.1007/978-3-030-42428-2_4

1   Introduction

Breast biopsy is the preferred technique to evaluate the malignancy of screening-detected suspicious lesions. To direct the needle towards the target, biopsy procedures are performed under image guidance, normally done with ultrasound (US) probes due to their ability to provide real-time visualization of both the needle and the internal structures [18]. However, proper needle placement with US remains a challenging task. First, malignant lesions cannot always be adequately visualized due to the poor image contrast of US. Furthermore, navigation towards complex 3D lesion geometries is commonly achieved using 2D freehand US (FUS) systems, which provide information in a lower-dimensional space [11]. Since highly sensitive pre-operative images (such as MRI or CT) can provide accurate positions of the lesions, finding a method to update these positions from real-time US images during an intervention would highly benefit current biopsy procedures. Several commercial and research platforms have implemented image fusion techniques that align pre-operative and intra-operative data, exploiting rigid or affine registration methods [6]. However, when dealing with breast anatomy, large deformations arise due to compression forces applied by the US probe. To provide accurate probe-tissue coupling and acceptable image quality, an appropriate alignment procedure of the pre-operative and US data is required.

Accurate modelling of soft tissue deformation in real-time is a far-from-being-solved problem. Biomechanical models relying on the finite element method (FEM) realistically calculate soft tissue deformations by using a mathematical model based on continuum mechanics theory. Although these models have been successfully employed for multimodal breast image registration, they have never been applied to registration between pre-operative data and intra-operative US, due to difficulties in providing a prediction within real-time constraints [8]. This is especially true when considering large, non-linear deformations which involve hyperelastic objects, as it is the case for the breast.

In order to meet real-time compliance, various techniques have been proposed to simplify the computational complexity of FEM. Some of them have focused on optimizing linear solvers (the main bottleneck of FEM) or the formulation itself, such as corotational [5] and multiplicative jacobian energy decomposition [13]. Very efficient implementations also exist, like Total Lagrangian explicit dynamics (TLED) [15], which can achieve real-time performances when coupled with explicit time integration and GPU-based solvers [10]. Another possible option to lower the simulation time is through dimensionality reduction techniques, like Proper Orthogonal Decomposition (POD), where the solution to a high-dimensional problem is encoded as a subset of precomputed modes. The most optimized approach used to model breast biomechanics is the one proposed by Han et al. in [7], which relies on GPU-based TLED formulation. Despite the significant simulation speedup achieved, solving the FE system took around 30 s, which is still not compatible with real-time. Modelling methods that do not rely on continuum mechanics laws have also been used to approximate soft tissues behavior. Among these, the position-based dynamics (PBD) approach has been used to predict breast lesions displacement due to US probe pressure in real-time, providing comparable accuracy with FE models [21]. However, not being based on real mechanical properties, such model requires an initial optimization of simulation parameters to obtain a realistic description of the deformation.

An emerging approach which has the potential of being both accurate and fast, exploits neural networks to estimate soft tissue behavior. Machine learning-based methods have proven successful to predict the entire 3D organ deformation starting either by applied surface forces [17, 22] or by acquired surface displacements [1, 19]. Being networks trained with synthetic data generated from FE simulations, they can reproduce a realistic physics-based description of the organ mechanical behavior. Using FE simulations for model training in the context of MRI-US deformable image registration has already been proposed in [9], where the authors build a statistical model of prostate motion which can account for different properties and boundary conditions. In the case of the breast, the potentiality of employing machine learning techniques has been already shown in [14], where several tree-based methods have been employed to estimate breast deformation due to compression between biopsy plates. These methods have been trained on 10 different patient geometries with a very specific FE simulation, where the upper plate is displaced vertically towards the lower one.

Similarly to works in [1, 19], we propose an approach where a neural network is trained to predict the deformation of internal breast tissues starting from the acquired surface displacements induced by the US probe. Our network can be seen as a patient-specific model. We train it on a single patient geometry before surgery, with a relatively small amount of training data. However, in contrast to the work of [14], FE simulations that compose the training set are generated with several random input displacements, making our approach able to generalize to different probe positions and compression extents.

The proposed method consists in a U-Net architecture, described in Sect. 2.2, and an immersed boundary method used for generating patient-specific simulations, described in Sect. 2.3. Results presented in Sect. 3 show the efficiency of the method when applied to both synthetic and ex vivo scenarios. Our contribution consists of a novel method to generate a real-time capable soft tissue model to improve target visualization during needle-based procedures. The position of lesions identified beforehand on pre-operative images can be updated from intra-operative ultrasound data and visualized by the surgeon in real-time.

2 Methods

This work presents a data-driven method to estimate in real-time the displacement of the breast internal structures due to probe pressure during US scanning. In our pipeline, we assume to have a patient-specific geometric model of the breast, obtained from pre-operative imaging such as MRI, and to know the position and orientation of the US probe at each time, thanks to a spatial tracking system. If the tracking coordinate system and the coordinate system of pre-operative imaging are registered, knowledge about the 3D pose and the geometry of the US probe directly allows to identify the contact surface between the breast and the probe. Since the US probe is represented as a rigid body, we can reasonably assume that when the anatomy is deformed by the probe during the image acquisition process, points on the breast surface below the US probe will be displaced to the same exact extent as the probe itself. As a consequence, our method can predict the displacements of all the points within the anatomy given as input the displacement of the surface nodes in contact with the US probe. The decision of relying on surface displacement inferred from the spatial tracking of the US probe instead of directly tracking surface deformations (through, for example, an RGBD camera) was taken from the fact that probe-induced deformations are large but local, and the probe itself would occlude most of the deformed surface to the sensor, thus preventing an accurate estimation of the contact surface displacements.

2.1 The U-Net Architecture

The objective of our work is to find the relation function f between the partial surface deformation under the US probe and the deformation inside the breast. Let $\mathbf{u_s}$ be the surface deformation and $\mathbf{u_v}$ the volumetric displacement field. In order to find f a minimization is performed on the expected error over a training set $\{(\mathbf{u_s}^n, \mathbf{u_v}^n)\}_{n=1}^{N}$ of N samples:

$$\min_{\theta} \frac{1}{N} \sum_{n=1}^{N} \| f(\mathbf{u_s}^n) - \mathbf{u_v}^n \|_2^2 \tag{1}$$

where θ is the set of parameters of the network f. We propose to use the same architecture as in [1], that is a U-Net [20] adapted to our application (see Fig. 1). The network consists of an encoding path that reduces the high dimensional input into a reduced space, and a decoding path that expands it back to the original shape. The skip connections transfer features along matching levels from the encoding path to the decoding path through crop and copy operations. As Fig. 1 shows, the encoding path consists of k sequences ($k = 3$ in our case) of two padded $3 \times 3 \times 3$ convolutions and a $2 \times 2 \times 2$ max pooling operation. At each step, each feature map doubles the number of channels and halves the spatial dimensions. In the lower part of the U-Net there are two extra $3 \times 3 \times 3$ convolutional layers leading to a 1024-dimensional array. In a symmetric manner, the decoding path consists of k sequences of an up-sampling $2 \times 2 \times 2$ transposed convolution followed by two padded $3 \times 3 \times 3$ convolutions. At each step of the decoding path, each feature map halves the number of channels and doubles the spatial dimensions. There is a final $1 \times 1 \times 1$ convolutional layer to transform the last feature map to the desired number

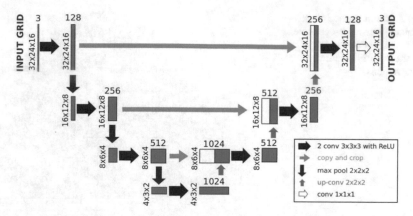

Fig. 1 U-Net architecture for a padded input grid of size $32 \times 24 \times 16$

of channels of the output (three channels in our case). The design of the U-Net is based on a grid-like structure due to this up- and down-sampling process. Hence we directly mesh our deformable object with regular hexahedral elements as explained in the next section.

2.2 Simulation of Breast Tissue Using Hexahedral Grids

The training data set consists of pairs of $(\mathbf{u}_s, \mathbf{u}_v)$ where $\mathbf{u}_s$ is the input partial surface displacement and $\mathbf{u}_v$ is the volumetric displacement field. Even though the data generation process takes place in an offline phase, in order to generate enough training data with FE simulations within clinically acceptable times (the intervention can be performed on the day after pre-operative scan is acquired), it is important to have simulations that are both accurate and computationally efficient.

We consider the boundary value problem of computing the deformation on a domain Ω under both Dirichlet and Neumann boundary conditions. Let Γ be the boundary of Ω (in our case, Γ corresponds to breast external surface, while Ω represents the entire breast volume). We assume that Dirichlet boundary conditions are applied to Γ_D and are a-priori known, whereas Neumann boundary conditions are applied to Γ_N, a subset of Γ that represents probe-tissue contact area and changes depending on current US probe position. In this work, training data for the network are generated by solving the discretized version of the following boundary value problem, exploiting the FE method:

$$\begin{cases} -\nabla \cdot \boldsymbol{\sigma} = \mathbf{0} \; in \; \Omega \\ \mathbf{u} = 0 \; on \; \Gamma_D \\ \boldsymbol{\sigma}\mathbf{n} = \mathbf{t} \; on \; \Gamma_N \end{cases} \tag{2}$$

Fig. 2 Breast surface mesh
obtained from a pre-operative
CT scan immersed in a
hexahedral grid for FEM
computations

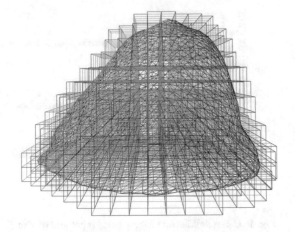

where σ is the Cauchy stress tensor, **n** is the unit normal to Γ_N and **t** is a traction force applied to the boundary. Note that in (2) we neglect all time-dependent terms and we do not apply any body force like gravity, since our geometric model already accounts for the effect of gravity force. The relation between stress and strain is described through the Saint Venant-Kirchhoff model, which is the simplest and most efficient extension of a linear elastic material to the nonlinear regime. This choice is motivated by the fact that a simple linear elastic model would not be able to appropriately describe the large deformations undergone by the breast. An iterative Newton-Raphson method is used to solve the non-linear system of equations approximating the unknown displacement.

We choose to discretize the domain into 8-node hexahedral elements not only for their good convergence properties and lock-free behavior, but also because it is the required structure for the input to the network. To do that, the 3D breast geometry is embedded in a regular grid of hexahedral elements (see Fig. 2) and we use an immersed-boundary method to correctly approximate the volume of the object in the FE method computations.

2.3 Data Generation

The input to the network corresponds to the displacement $\mathbf{u}_s$ of the points belonging to the breast-probe contact area. The punctual displacements are spread to the nodes of the surrounding cuboid cell through a barycentric mapping and the corresponding volume displacement $\mathbf{u}_v$ is obtained by the previously explained FE approach in response to $\mathbf{u}_s$. The data used to train the network must be representative of the application scenario and must allow the network to extract the pertinent features of the tissue behavior. In order to train our model to estimate breast volume deformation in response to pressure imposed with the US probe, we simulate several random probe-induced deformations using the following strategy:

- Select a random node p in the breast surface
- Select an oriented bounding box A centered in point p and normal to the breast surface, whose dimensions match those of the US probe lower surface, which represents current probe-tissue contact area
- Select all the surface points P falling within the box A
- Select as force direction d the normal to the surface at point p plus a random angle α ($\alpha \in \left[-\frac{\pi}{4}, \frac{\pi}{4}\right]$)
- Apply the same force f of random magnitude ($|f| \in [0.0, 0.8]$) along direction d to the P selected points simultaneously
- Store the displacement at the set of points P (input to the network) and the displacement of all the points in the volume (output to the network)
- Repeat the procedure until $N + M$ samples are generated

The choice of applying force f allowing some angle deviation from normal direction enables us to include in our dataset samples where the probe compression is not precisely normal to the surface, as it can be the case in freehand US acquisitions. The maximal force magnitude (e.g., $0.8\,N$) is set such that the amount of maximal deformation reproduced in the training dataset never exceeds too much that observed in real clinical settings. The described strategy is used to generate the set $\{(\mathbf{u_s}^n, \mathbf{u_v}^n)\}_{n=1}^{N}$ of N samples which is used to train the network, and the set $\{(\mathbf{u_s}^n, \mathbf{u_v}^n)\}_{n=1}^{M}$ of M samples which is left for validation. The training dataset is generated with the SOFA framework [3] on a laptop equipped with an Intel i7-8750H processor and 16 GB RAM.

3 Experiments and Results

The network presented in this work is used to predict US probe-induced deformations of a realistic multi-modality breast phantom (Model 073; CIRS, Norfolk, VA, USA). The 3D geometry model of the phantom surface and 10 inner lesions (diameter of 5–10 mm) is obtained by segmenting the corresponding CT image, relying on ITK-SNAP and MeshLab frameworks [2, 24]. A Freehand Ultrasound System (FUS) based on a Telemed MicrUs US device (Telemed, Vilnius, Lithuania) equipped with a linear probe (model L12-5N40) is used to acquire US images of the 10 segmented lesions. The dimension of the probe surface is (5×1 cm). For each lesion, we acquire US images in correspondence of four different input deformations. The MicronTracker H×40 (ClaronNav, Toronto, Canada) optical tracking system is used to track US probe in space (Fig. 3a). The overall probe spatial calibration error is below 1 mm (± 0.7147), estimated through the PLUS toolkit [12]. Landmark-based rigid registration is performed to refer the CT-extracted 3D model, the US probe and the US images to the same common coordinate system, exploiting 3D Slicer functionalities [4]. The registration process does not only enable us to extract the breast-probe contact area, as described in

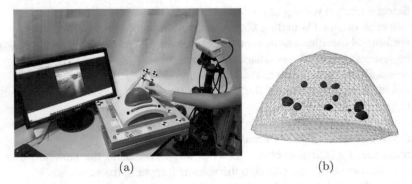

(a)   (b)

Fig. 3 (**a**) Experimental setup. From left to right: monitor showing real-time US images; CIRS breast phantom during FUS acquisition; optical tracking system that allows to map the real positions of the CIRS breast phantom and the US probe to the preoperative geometry model. (**b**) External surface and inner lesions of the CIRS breast phantom

Sect. 2, but also to know in real-time the 3D position of any point belonging to the US image. In this way, it is possible to refer lesions position extracted from US images to the 3D space.

3.1 Predict Displacement on Synthetic Data Sets

Elastic properties of the physics model used to generate training data are set in accordance with the values estimated in [23] for the same breast phantom considered in this study. However, as we are imposing surface displacements, the values of the elasticity parameters do not affect the displacement field inside the simulated volume as long as the ratio of the different stiffness values is maintained [16], thus making the method reliable for any patient specificity. Dirichlet boundary conditions are imposed by constraining the motion of all the nodes belonging to the lowest phantom surface.

Using the method described in Sects. 2.2 and 2.3, we discretized the breast phantom into 2174 hexahedral elements and we simulated several probe-induced displacements. Overall we generated $N = 800$ samples for training and $M = 200$ samples for testing. The U-Net is trained in a GeForce GTX 1080 Ti using a batch size of 4, 100,000 iterations and the Adam optimizer. We used a Pytorch implementation of the U-Net. To assess the learning capability of the network, we perform a statistical analysis of the *mean norm error e* over the testing data set. Let $\mathbf{u_v}^m$ be the ground truth displacement tensor for sample m generated using the finite element method described in Sect. 2.2 and $f(\mathbf{u_s}^m)$ the U-Net prediction. The mean norm error between $\mathbf{u_v}^m$ and $f(\mathbf{u_s}^m)$ for sample m reads as:

Table 1 Error measures over the testing data set for a breast having 2174 H8 elements, with maximal nodal deformation of 79.09 mm

$\bar{e}$ (mm)	$\sigma(e)$ (mm)	$\max\limits_{m \in M} e$ (mm)	Prediction time (ms)	Total training time (min)
0.052	0.050	0.266	3.14 ± 0.56	278

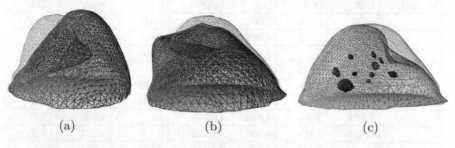

(a)   (b)   (c)

Fig. 4 (**a**) Sample with maximal deformation (79.09 mm). (**b**) Sample with maximal *mean norm error* (0.266 mm). The green mesh is the U-Net prediction and the red mesh is the FEM solution. The initial rest shape is shown in grey. (**c**) U-Net prediction on phantom data

$$e(\mathbf{u_v}^m, f(\mathbf{u_s}^m)) = \frac{1}{n} \sum_{i=1}^{n} |\mathbf{u_{v_i}}^m - f(\mathbf{u_s}^m)_i|. \tag{3}$$

where n is the number of degrees of freedom of the mesh. We compute the average $\bar{e}$, standard deviation $\sigma(e)$ and maximal value of such norm over the testing data set. The obtained results are shown in Table 1. The maximal error is of only 0.266 mm and corresponds to the sample shown in Fig. 4b. The most striking result is the small computation time required to make the predictions: only 3.14 ± 0.56 ms. In contrast, the FE method takes on average 407.7 ± 64 ms to produce the solution. Obviously, the resolution of the FE mesh could be reduced to accelerate the computations but at the cost of an accuracy loss.

3.2 Predict Displacement on Phantom Data

In our experiments, we consider one lesion at a time and we reposition the US probe on the surface of the breast such that the lesion considered is visible on the US image. In order to validate our model, we manually extract lesions position from US image acquired at rest (i.e., without applying any deformation, when the probe is only slightly touching the surface) and we consider it as a landmark to track. We then impose four deformations of increasing extent for each lesion, and we compare the U-Net-predicted displacement with real displacements extracted from US images. The comparison is performed computing target registration error (TRE) between the predicted position of the lesion and its ground-truth position. The performance of

Table 2 Target registration errors in millimeters for different tumors and different deformation ranges in the breast phantom. The first table is for the proposed method, while the second table reports results obtained with the FE model used for data generation. Not-acquired data is reported as (−)

U-Net predictions							
TumorID	D10	D15	D20	D25	D30	Mean	STD
1	–	1.936	2.002	1.506	3.053	2.124	0.569
2	3.211	2.905	4.068	–	4.137	3.580	0.534
3	2.032	–	4.709	7.134	10.90	6.194	3.262
4	0.505	2.225	5.313	5.903	–	3.486	2.217
5	0.932	2.768	3.454	–	4.893	3.012	1.425
6	3.923	6.349	5.625	–	6.724	5.655	1.075
7	3.454	3.864	4.543	6.710	–	4.643	1.255
8	2.422	3.261	4.320	5.136	–	3.785	1.030
9	–	3.928	4.214	4.578	4.858	4.394	0.353
10	5.529	3.272	3.940	4.846	–	4.397	0.860
Mean	2.751	3.390	4.219	5.116	5.761		
STD	1.638	1.294	1.007	1.854	2.788		
FE method							
TumorID	D10	D15	D20	D25	D30	Mean	STD
1	–	1.326	2.151	2.075	3.759	2.328	0.887
2	1.956	2.738	3.945	–	4.025	3.166	0.865
3	1.595	–	4.748	7.044	10.932	6.080	3.404
4	0.755	1.991	4.544	5.120	–	3.103	1.795
5	1.029	2.863	3.330	–	4.541	2.941	1.262
6	2.579	3.409	2.871	–	2.337	2.799	0.400
7	2.605	3.219	4.095	6.750	–	4.167	1.582
8	2.695	2.748	4.321	5.411	–	3.794	1.139
9	–	2.745	2.497	2.510	4.193	2.986	0.704
10	2.916	2.542	3.015	3.868	–	3.085	0.485
Mean	2.016	2.620	3.552	4.682	4.964		
STD	0.765	0.593	0.856	1.803	2.757		

our method is compared to that of the FE model used for data generation. In Table 2 are shown the target registration errors for each phantom lesion with respect to the applied deformation. The input deformations are classified into five ranges based on the probe displacements. Displacement ranges indicated as D15, D20 and D25 have a fixed length of 5 mm each and are centered at 15, 20 and 25 mm respectively. D10 and D30 contain the extreme cases under 12.5 mm or above 27.5 mm.

Values in Table 2 highlight that the average TRE for all the tumors and for all the deformations is smaller than 6.194 mm which is comparable to the maximum value obtained with the FE method (6.080 mm). The average error increases with the deformation range just like in the FE method. There is no significant difference between the values of the two tables, meaning that in terms of accuracy, our method is comparable to the data generation method used to train it. In order to compute each deformation, the FE method needs about 407.7 ms whereas the U-Net predicts the deformation in only 3 ms.

4 Conclusion

In this work we have proposed to use a deep neural network to learn the deformable behavior of the breast from numerical simulations based on the finite element method, in order to bypass the high computational cost of the FEM. Our approach represents an interface between precise biomechanical FE modeling (not capable of real time) and clinical applications requiring both high accuracy and very high speed. We have shown that our framework allows for extremely fast predictions of US probe-induced displacements of the breast during US scanning, achieving comparable accuracy to other existing methods. Therefore, it has the potential to be employed to update in real-time the estimated position of breast lesions identified on a pre-operative scan on US images, enabling continuous visualization of the biopsy target, even when sonography fails to render it.

Although the FE model used to train our network does not perform in real-time, its prediction delay of less than 1 s might be considered already acceptable for our specific application. However, such good computational performance is achieved since in this preliminary evaluation we use a very simplistic model, that does not account for heterogeneity or complex boundary conditions happening in clinical cases. Usage of a more complex FE model will certainly cause an increase of computation load. On the contrary, an important feature of our approach is that the prediction time remains close to 3 ms regardless of the grid resolution and of the biomechanical model used for the data generation process. This means that increasing the complexity of the model used to generate the data set will not affect the prediction speed. Moreover, our pipeline allows the method to be insensitive to patient specific elastic properties as it imposes surface displacements. It is worth noting that for inhomogeneous objects, the displacement field still depends on the ratio of the different stiffnesses [16]. Another advantage of our method is the easy meshing process. Any geometry can be embedded in a sparse grid and through the use of immersed boundary simulations the deformations are correctly estimated.

The main limitation of our method remains the training process, which is burdensome and has to be repeated for every new geometry or application. However, we have shown that a limited amount of training data can be sufficient to train a U-Net such that it obtains accurate prediction within clinically acceptable times. As a future work, we plan to use a more general training strategy leading to a network model able to predict deformations induced by any type and number of compression tools (for example, different probe shapes or the two biopsy compression plates).

References

1. J.N. Brunet, A. Mendizabal, A. Petit, N. Golse, E. Vibert, S. Cotin, Physics-based deep neural network for augmented reality during liver surgery. MICCAI (2019)
2. P. Cignoni, M. Callieri, M., Corsini, M. Dellepiane, F. Ganovelli, G. Ranzuglia, in *MeshLab: An Open-Source Mesh Processing Tool*, ed. by V. Scarano, R.D. Chiara, U. Erra. Eurographics Italian Chapter Conference, The Eurographics Association (2008)
3. F. Faure, C. Duriez, H. Delingette et al., in *Sofa: A Multi-model Framework for Interactive Physical Simulation*, Soft Tissue Biomechanical Modeling for Computer Assisted Surgery (Springer, 2012), pp. 283–321
4. A. Fedorov, R. Beichel, J. Kalpathy-Cramer, J. Finet, J.C. Fillion-Robin, S. Pujol, C. Bauer, D. Jennings, F. Fennessy, M. Sonka, J. Buatti, S. Aylward, J. Miller, S. Pieper, R. Kikinis, 3D slicer as an image computing platform for the quantitative imaging network. J. Magn. Reson. Imaging **30**(9), 1323–1341 (2012)
5. C.A. Felippa, B. Haugen, A unified formulation of small-strain corotational finite elements: I. theory. Comput. Methods Appl. Mech. Eng. **194**(21–24), 2285–2335 (2005)
6. R. Guo, G. Lu, B. Qin, B. Fei, Ultrasound imaging technologies for breast cancer detection and management: a review. Ultrasound Med. Biol. **44**, 37–70 (2017)
7. L. Han, J.H. Hipwell, B. Eiben et al., A nonlinear biomechanical model based registration method for aligning prone and supine MR breast images. IEEE Trans. Med. Imaging **33**(3), 682–694 (2013)
8. J.H. Hipwell, V. Vavourakis, L. Han et al., A review of biomechanically informed breast image registration. Phys. Med. Biol. **61**(2), R1 (2016). http://stacks.iop.org/0031-9155/61/i=2/a=R1
9. Y. Hu, H.U. Ahmed, Z. Taylor, C. Allen, M. Emberton, D. Hawkes, D. Barratt, MR to ultrasound registration for image-guided prostate interventions. Med. Image Anal. **16**(3), 687–703 (2012)
10. G.R. Joldes, A. Wittek, K. Miller, Real-time nonlinear finite element computations on GPU– application to neurosurgical simulation. Comput. Methods Appl. Mech. Eng. **199**(49–52), 3305–3314 (2010)
11. J. Krücker, S. Xu, A. Venkatesan, J.K. Locklin, H. Amalou, N. Glossop, B.J. Wood, Clinical utility of real-time fusion guidance for biopsy and ablation. J. Vasc. Interv. Radiol. **22**(4), 515–524 (2011)
12. A. Lasso, T. Heffter, A. Rankin, C. Pinter, T. Ungi, G. Fichtinger, Plus: open-source toolkit for ultrasound-guided intervention systems. IEEE Trans. Biomed. Eng. **61**(10), 2527–2537 (2014)
13. S. Marchesseau, T. Heimann, S. Chatelin, R. Willinger, H. Delingette, in *Multiplicative Jacobian Energy Decomposition Method for Fast Porous Visco-hyperelastic Soft Tissue Model*, MICCAI (Springer, 2010), pp. 235–242
14. F. Martínez-Martínez, M.J. Rupérez-Moreno, M. Martínez-Sober et al., A finite element-based machine learning approach for modeling the mechanical behavior of the breast tissues under compression in real-time. Comput. Biol. Med. **90**, 116–124 (2017)
15. K. Miller, G. Joldes, D. Lance, A. Wittek, Total lagrangian explicit dynamics finite element algorithm for computing soft tissue deformation. Commun. Numer. Methods Eng. **23**(2), 121–134 (2007)
16. K. Miller, J. Lu, On the prospect of patient-specific biomechanics without patient-specific properties of tissues. J. Mech. Behav. Biomed. Mater. **27**, 154–166 (2013)
17. K. Morooka, X. Chen, R. Kurazume, S. Uchida, K. Hara, Y. Iwashita, M. Hashizume, in *Real-time Nonlinear FEM with Neural Network for Simulating Soft Organ Model Deformation*, MICCAI (Springer, 2008), pp. 742–749
18. E. O'Flynn, A. Wilson, M. Michell, Image-guided breast biopsy: state-of-the-art. Clin. Radiol. **65**(4), 259–270 (2010)
19. M. Pfeiffer, C. Riediger, J. Weitz, S. Speidel, Learning soft tissue behavior of organs for surgical navigation with convolutional neural networks. Int. J. Comput. Assist. Radiol. Surg. **14**, 1147–1155 (2019)

20. O. Ronneberger, P. Fischer, T. Brox, U-net: convolutional networks for biomedical image segmentation. MICCAI **9351**, 234–241 (2015)
21. E. Tagliabue, D. Dall'Alba, E. Magnabosco, C. Tenga, I. Peterlik, P. Fiorini, Position-based modeling of lesion displacement in ultrasound-guided breast biopsy. Int. J. Comput. Assist. Radiol. Surg. **14**, 1329–1339 (2019)
22. M. Tonutti, G. Gras, G.Z. Yang, A machine learning approach for real-time modelling of tissue deformation in image-guided neurosurgery. Artif. Intell. Med. **80**, 39–47 (2017)
23. F. Visentin, V. Groenhuis, B. Maris et al., Iterative simulations to estimate the elastic properties from a series of MRI images followed by MRI-us validation. Med. Biol. Eng. Comput. **194**(21–24), 1–12 (2018)
24. P.A. Yushkevich, J. Piven, H. Cody Hazlett et al., User-guided 3D active contour segmentation of anatomical structures: significantly improved efficiency and reliability. Neuroimage **31**(3), 1116–1128 (2006)

An Improved Coarse-Grained Model to Accurately Predict Red Blood Cell Morphology and Deformability

Nadeeshani Maheshika Geekiyanage, Robert Flower, Yuan Tong Gu, and Emilie Sauret

Abstract Accurate modelling of red blood cells (RBCs) has greater potential over experiments, as it can be more robust and significantly cheaper than equivalent experimental procedures to investigate the mechanical properties, rheology and dynamics of RBCs. The recent advances in numerical modelling techniques for RBC studies are reviewed in this study, and in particular, the discrete models for a triangulated surface to represent the in-plane stretching energy and out-of-plane bending energy of the RBC membrane are discussed. In addition, an improved RBC membrane model is presented based on coarse-grained (CG) technique that accurately and efficiently predicts the morphology and deformability of a RBC. The CG-RBC membrane model predicts the minimum energy configuration of the RBC from the competition between the in-plane stretching energy of the cytoskeleton and the out-of-plane bending energy of the lipid-bilayer under the given reference states of the cell surface area and volume. A quantitative evaluation of several cellular measurements including length, thickness and shape factor, is presented between the CG-RBC membrane model and three-dimensional (3D) confocal microscopy imaging generated RBC shapes at equivalent reference states. The CG-RBC membrane model predicts agreeable deformation characteristics of a healthy RBC with the analogous experimental observations corresponding to optical tweezers stretching deformations. The numerical approach presented here forms the foundation for investigations into RBC morphology and deformability under diverse shape-transforming scenarios, in vitro RBC storage, microvascular circulation and flow through microfluidic devices.

Keywords Coarse-graining · Deformability · Discocyte · Elongation index · Morphology · Morphology index · Numerical modelling · Optical tweezers stretching · Red blood cell · Shape factor

N. M. Geekiyanage · Y. T. Gu · E. Sauret (✉)
School of Mechanical, Medical and Process Engineering, Queensland University
of Technology (QUT), Brisbane, QLD, Australia
e-mail: emilie.sauret@qut.edu.au

R. Flower
Research & Development, Red Cross Lifeblood, Brisbane, QLD, Australia

K. Miller et al. (eds.), *Computational Biomechanics for Medicine*,
https://doi.org/10.1007/978-3-030-42428-2_5

47

Abbreviations

2D	Two-dimensional
3D	Three-dimensional
ADE	Area-difference-elasticity
AFM	Atomic force microscopy
BCM	Bilayer-coupling model
BIM	Boundary integral method
CG	Coarse-graining
CGMD	Coarse-grained molecular dynamics
DPD	Dissipative particle dynamics
FEM	Finite element method
HE	Hereditary elliptocytosis
HPC	High performance computing
HS	Hereditary spherocytosis
IBM	Immersed boundary method
MD	Molecular dynamics
QUT	Queensland University of Technology
RBC	Red blood cell
SAGM	Saline-adenine-glucose-mannitol
SCM	Spontaneous curvature model
SEM	Scanning electron microscopy
SF	Shape factor
SP	Spring-particle
SPH	Smoothed particle hydrodynamics
TEM	Transmission electron microscopy
WLC	Worm-like-chain

1   Introduction

Red blood cells (RBCs), though remarkably simple in structure [1], perform a vital physiological function, transferring oxygen and carbon dioxide between lung and body tissues. RBCs are composed of a composite membrane surrounding a haemoglobin rich cytoplasm, and the cell deformability is primarily influenced by mechanical and geometrical factors of the cell such as cell surface area and volume, elasticity and viscosity of the cell membrane, and volume and viscosity of the cytosol [2–8]. The changes in the cell membrane structure and its mechanical properties adversely affect the cell deformability, and the loss of cell deformability is a potential indicator of cell functional impairments in many pathophysiological conditions [9–12]. Therefore, loss of RBC deformability is an indicator of cell functional impairments in many pathophysiological conditions [9–12]. The RBC deformability and its morphology are associated together, and changes to the

Fig. 1 Representation of the cross-sectional view of a biconcave shape of a healthy RBC

healthy biconcave discocyte morphology reflect the impaired cell deformability. The RBC morphology is influenced by the cell age, in several diseased conditions (e.g. hereditary spherocytosis, hereditary elliptocytosis, and sickle cell anaemia) [13] and some extracellular environmental conditions (e.g. amphiphilic substances, osmolality, ionic strength and pH) [14].

Different experimental and numerical techniques have been applied to investigate the physical, mechanical, rheological, and dynamic properties of RBCs under a variety of healthy and diseased conditions. In particular, numerical modelling is an attractive approach to overcome some of the experimentation-related challenges and have been applied successfully to investigate the RBC morphology and deformability in a variety of circumstances. In this background, this chapter initially presents an overview of recent advances in numerical modelling techniques for RBC studies. Especially, the discrete models for a triangulated surface to represent the in-plane stretching energy and out-of-plane bending energy of the RBC membrane are discussed. Then, an improved RBC membrane model is presented based on coarse-grained (CG) technique that accurately and efficiently predicts the morphology and deformability of a RBC, which is followed by a comprehensive discussion of its key applications, limitations and future prospects.

1.1 RBC Cellular Structure

RBCs not only play the critical role of transporting oxygen and carbon dioxide between lungs and body tissues [15], but are also involved in inflammatory processes and coagulation [16]. RBCs are unique nucleus-free cells [16–19], of which 95% of its cytoplasm is haemoglobin, the metalloprotein responsible for oxygen transfer [16, 20]. A healthy RBC at physiological conditions assumes the shape of a biconcave disc with dimensions of $\sim 8\ \mu m$ in diameter [2, 21, 22] and $\sim 2\ \mu m$ in thickness [20], and a simple representation of a healthy RBC is presented in Fig. 1. Having a cell volume of $\sim 90\ \mu m^3$ and cell surface area of $\sim 140\ \mu m^2$, a RBC holds 40% excess surface area compared to a sphere with the same volume [19, 23]. The RBC cell membrane consists of a fluid bilayer of thickness $\sim 4\ nm$ [21] and a cytoskeletal complex [16, 17, 20, 24]. The offset between the bilayer and the cytoskeleton is only 30–50 nm [21, 25].

The bilayer is a complex combination of phospholipids, cholesterol and dissolved proteins that are asymmetrically distributed in the two leaflets of the bilayer [26], and acts as an osmotic barrier for the cell controlling the passage of water, ions and larger solute molecules through it [21]. The major constituents of the outer bilayer-leaflet are lipids such as phosphatidylcholine, sphingomyelin, and glycophospholipids, whereas phosphatidylserine and phosphatidylethanolamine are the major constituents of the inner leaflet [21, 27]. The predominant lipids in the outer leaflet are neutral at physiological pH, whereas the phosphatidylserine in the inner leaflet is negatively charged, and therefore, there exists a significant charge difference between these two leaflets. The hydrophobic ends of the phospholipids are embedded in the bilayer while the hydrophilic ends are protruding from it.

The thin cytoskeleton is a hexagonally linked network which is composed primarily of spectrin filaments and actin protofilaments [28]. Each spectrin tetramer consists of two heterodimers of intertwined and antiparallel α-spectrin and β-spectrin filaments with an extended length of $\sim$ 200 nm [29]. There are about 33,000 hexagonal junctional complex structures in the RBC cytoskeleton, of which the actin protofilament is the central piece while up to six spectrin dimers are connected to this. The distance between the vertices of the hexagonal junctional complex is $\sim$ 76 nm. The head-to-head association of spectrin dimers that connects them into a tetramer links these junctional complexes horizontally, and there are vertical connections to link the cytoskeleton with the bilayer as well. The cytoskeleton-bilayer connection through ankyrin, protein 4.2 and band 3 protein is the primary, whereas the connection at actin protofilaments through protein 4.1 and glycophorin C is the secondary connection [14, 17, 19, 24, 30]. The ability of band 3 and glycophorin C to drift within the lipid-bilayer provides horizontal mobility to the bilayer-cytoskeleton connection [28]. Although the RBC membrane is heterogeneous at the molecular length scale, it can reasonably be approximated to be homogeneous in its properties on scales above 100 nm [4, 21]. The following subsection discusses the morphology of a healthy RBC and some morphology transformation conditions where different RBC morphologies can be observed.

1.2 RBC Morphology

RBCs display a range of morphologies corresponding to the cell age, several diseased conditions, and some extracellular environmental conditions. The characteristic biconcave shape of a healthy RBC observed during normal physiological conditions is acknowledged as the 'discocyte' morphology. The discocyte morphology of a healthy RBC gradually transforms into echinocyte and then to sphero-echinocyte during storage.

RBC diseases, such as hereditary spherocytosis (HS), hereditary elliptocytosis (HE) and sickle cell disease, alter the RBC morphology as well [5, 13, 19, 23, 30–34]. RBC diseases are often associated with defects from mutations in proteins modifying their inter-connectivity and their connectivity with the lipid-bilayer [30]. For example, under HS and HE conditions RBCs become spherical due to

Fig. 2 Scanning electron microscope (SEM) images of (**a**) stomatocyte, (**b**) discocyte and (**c**) echinocyte RBC morphologies (M.A. Balanant, unpublished data, personal communication, March 20, 2018)

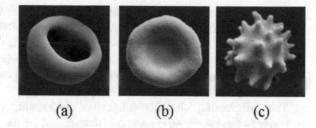

(a)   (b)   (c)

partial loss of the lipid-bilayer and contain higher haemoglobin concentration than usual; malaria infection alters the RBC membrane properties and make the cell spherical at the later stages of parasite development; and sickle cell anaemia causes haemoglobin polymerization [5, 13, 19, 23, 30–34]. In addition, modifications to the extracellular environment can influence the RBC morphology, and in particular, there are stomatocytogenic and echinocytogenic shape-transforming environments that can produce stomatocytes and echinocytes respectively [14, 18, 19, 27, 35–53]. The appearance of the 'stomatocyte' and 'echinocyte' morphologies are cup-shaped and spiculated respectively, and representative images of stomatocyte, discocyte and echinocyte morphologies are presented in Fig. 2. A detailed classification of many RBC morphologies along with probable disease conditions and extracellular environmental factors is first presented by Bessis [52]. For example, the RBC morphology becomes stomatocytic and echinocytic at cellular pH levels of 5.6 and 8.8, respectively [43]. In addition, echinocyte forms can be observed at 300 mmHg on day 5 of storage in phosphate-dextrose-adenine-1 preservative solution [40]; due to the adhesion of nano-diamonds on RBC membrane [39]; and under storage in saline-adenine-glucose-mannitol (SAGM) at days 42 of storage [54]. A detailed classification of many RBC morphologies along with probable disease conditions and extracellular environmental factors is first presented by Bessis [52]. The RBC morphology and its deformability are interlinked, and the following subsection details the deformability determinants of a RBC.

1.3   RBC Deformability

RBCs require high deformability to sustain passage through the microcirculation [17, 55]. The determinants of the cell deformability are the cell geometry, the cytoplasmic viscosity, and the membrane deformability characteristics [3, 8, 13, 22, 23, 56]. The cell geometry can be attributed to the ratio of the cell surface area to its volume. The larger cell surface area not only increases the exchange of gases that take place on its surface [57], but also facilitates a broad range of RBC shapes under a variety of mechanical deformation scenarios. However, the surface area incompressibility of the lipid-bilayer, shear resistance of the cytoskeleton and the volumetric incompressibility of the cytosol, limit any change in cell geometry.

The cohesion between the cytoskeleton and the bilayer facilitate the maintenance of the cell surface area under deformation. A healthy RBC can deform with linear extensions of up to $\sim 250\%$, however, a 3–4% increase in surface area results in cell lysis [19]. Loss of cell surface area due to membrane vesiculation and cell fragmentation, and any change in cell volume due to defective ion transport [19, 58] adversely influences RBC deformability.

The cytoplasmic viscosity influences the rapidity of cell shape changes in response to fluid shear stresses and is determined by the intracellular haemoglobin concentration. The maintenance of cell haemoglobin concentration within a very narrow range minimizes the fluctuations in cytoplasmic viscosity [19, 56, 59], and conserves the cell flexibility to traverse narrow capillaries in the microcirculation. The RBC membrane is highly deformable and facilitates large reversible deformation of the cell. The lipid-bilayer contributes to the cell's bending resistance, whereas the cytoskeletal spectrin network contributes to its shear resistance [13, 19, 22, 23, 60, 61]. The bending resistance is characterized by the bending modulus (κ), and κ for a healthy RBC lies between 1×10^{-19} and 7×10^{-19} Nm [13]. Similarly, the membrane shear resistance is characterized by the shear modulus (μ_0), and μ_0 for a healthy RBC lies between 4 and 12 μNm^{-1} [13, 62, 63]. The structural organization of the RBC membrane is crucial for the cell to maintain its shape and mechanical integrity [16]. The dynamic equilibrium of the cell membrane exists due to the association and dissociation of the inter-protein and intra-protein linkages. It is possible to rupture the cytoskeletal connections and cytoskeleton-bilayer connections under mechanical loads [28], which can lead to structural instability and lowered deformability of the RBC.

The RBC deformability characteristics are strongly linked to structural and molecular alterations induced by the onset and progression of many pathophysiological conditions [2, 64–66]. Altered lipid composition and bilayer asymmetry that influences the RBC shape, and the modifications in cytoskeletal proteins that affect the RBC membrane integrity, have severe implications on the RBC function and viability [17]. Less deformable RBCs can obstruct capillaries and require significantly higher transit time to navigate through the microvasculature, leading to decreased levels of oxygen delivery to organs [67–69]. In addition, less deformable RBCs are promptly removed from the circulation at the spleen [70, 71]. Therefore, RBC deformability is a potential measure of cell functional impairments under many pathophysiological conditions [65]. As such, the associated changes in RBC membrane elasticity, cell geometry and intracellular viscosity during RBC morphology transformation corresponding to cell age, disease conditions, and extra-cellular environment, affect the cell deformability. Therefore, the RBC morphology and its deformability are linked together, and the deviation of RBC morphology from the discocyte shape towards other RBC morphologies generally indicates impaired cell deformability [3, 25, 72, 73]. The measurements of membrane shear modulus and bending modulus of a discocyte indicate significant increase during RBC morphology transformations as well [72, 74]. However, the change in RBC deformability is associated with the stage of its morphology rather than the severity of the morphology transformation condition [3].

Numerous experimental and numerical studies have been performed to investigate the physical, mechanical, rheological, and dynamic properties of RBCs under a variety of physiological and pathophysiological circumstances (e.g. membrane vesiculation, membrane defects, cell lysis, optical tweezers stretching, micropipette aspiration, atomic force microscopy (AFM) indentation, membrane thermal fluctuations, large-scale blood flow, cell margination, and microfluidics) [4, 6, 13, 20, 22, 23, 63, 64, 72, 74–85]. However, some experimental and pre-preparation procedures can influence the associated measurements. For example, the limited resolution of bright field and phase contrast microscopy imaging, require cell fixation prior to SEM imaging, cell adhesion on the substrate prior to AFM studies, and cell proximity to glass surface during experiments can influence the cell response, and therefore, the exact measurements [86]. It is difficult to control precisely the exact location of the RBC membrane-bead attachment under optical tweezers technique, which can affect the cell response [87]. In addition, certain experimental techniques are time consuming, labour-intensive, require special technical skills, expensive equipment, and may require high-speed video microscopy systems as well (e.g. optical tweezers, micropipette aspiration, AFM, and microfluidics) [75]. The cell response to experimental techniques is inconsistent from donor to donor. As a result, most of the measured cell parameters are averaged values based on these inhomogeneous RBC samples. However, the selection of most appropriate technique that matches with the experimental objectives results in a better determination of RBC properties. In addition, numerical analysis established on experimental studies can facilitate more efficient and effective decision-making on RBC systems. Following subsection reviews recent advances in numerical modelling techniques for RBC studies.

1.4   Numerical Investigations on RBC Morphology and Deformability

Accurate modelling of RBCs has great potential, as it can be more robust and significantly cheaper than the equivalent experimental procedures [84, 88]. For example, the experimental outcomes can be influenced by factors such as the experimental protocol in use, any experimental errors or uncertainties and donor variability, and lead to a statistically significant deviation of the results. However, accurate RBC modelling can produce consistent predictions. In addition, the rapid advancement in computational systems enables very large and sophisticated simulations analogous to complex experimental techniques. Due to the simplicity of the RBC structure, it can be numerically approximated as a bag of concentrated haemoglobin solution surrounded by a thin macroscopically homogeneous membrane [21]. The RBC membrane can be treated as a two-dimensional (2D) viscoelastic surface in a three-dimensional (3D) space in the cellular length scale [13, 20, 21], as the thickness of the lipid-bilayer is only $\sim$ 4 nm [18] and the offset between lipid-bilayer and cytoskeleton is only $\sim$ 30–40 nm [18]. In addition, the heterogeneous nature of the RBC membrane can be reasonably approximated to be homogeneous in its properties for length scales above 100 nm [21, 89] and is suitable for mesoscopic and macroscopic scale investigations. However, a more realistic and accurate RBC representation requires a detailed description of its structure, especially for studies on the mechanics of many pathophysiological conditions [30], and therefore, necessitates the incorporation of the properties of the lipid-bilayer, cytoskeleton, transmembrane proteins and their interrelation.

There are several types of numerical modelling techniques for RBC studies, and the primary approaches are continuum, particle-based and hybrid continuum-particle [5, 20, 80, 89] based techniques. Continuum-based numerical techniques (e.g. finite element method (FEM), boundary integral method (BIM), and immersed boundary method (IBM)) treat the RBC membrane and associated fluid components as homogeneous materials, whereas the particle-based numerical techniques (e.g. dissipative particle dynamics (DPD), smoothed particle hydrodynamics (SPH), lattice Boltzmann method (LBM), molecular dynamics (MD), and coarse-grained molecular dynamics (CGMD)) represent these components via a network of particulate assembly [5, 13, 20, 31, 32, 63, 84, 90]. Continuum-based modelling has been successfully applied to study large-scale blood flow conditions [85], tank-treading motion of RBCs [91], and optical tweezers stretching deformation [2]. Continuum-based modelling can accurately predict the RBC behaviour on the whole cell level but has limited potential to capture the subcellular and molecular level details, whereas particle-based modelling can successfully capture these details. In addition, it is easier to implement complex structures with particle-based modelling, since this method is based on arbitrarily distributed particles [80]. DPD and LBM techniques can take into account the thermal fluctuations, and therefore are suitable for studies where thermal fluctuations play a significant role (e.g. RBCs aggregation) [80]. In addition, DPD and SPH techniques consider a set of particles inside a specified influence domain, and therefore cause a lower computational cost than LBM, which is a hybrid-mesh particle based method.

There are molecularly detailed RBC membrane models, which successfully capture the membrane structure and its response under both normal and defective states [1, 5, 7, 13, 22, 62, 63, 92–94]. However, these models can be computationally very expensive when extended for large systems containing multiple cells, and therefore, coarse-grained techniques and/or hybrid continuum-particle-based techniques are probable solutions to reduce the computational complexity [5, 13, 20, 29, 62, 63, 81, 84, 90, 95–98]. Coarse-grained (CG) particles represent a cluster of particles, and capture the important features with a smaller number of particles while effectively reducing the computational expense [29, 97, 98]. For example, the two-component CGMD composite model developed by Li and Lykotrafitis [33] investigated the mechanisms of RBC membrane vesiculation for a small piece of the membrane of 0.8 μm $\times$ 0.8 μm and is composed of 32,796 CG particles. The OpenRBC model by Tang et al. [92] represents the RBC membrane to the level of protein resolution and facilitates studies such as RBC vesiculation and lysis under different pathophysiological conditions. OpenRBC is strengthened with many features to optimize simulation efficiency (i.e. adaptive partitioning of the particles, parallelization, and simultaneous hardware threads). For example, OpenRBC can simulate a whole RBC composed of 3,200,000 particles in only 1346 s on IBM POWER8 "Minsky" computer system through 1 core and 1 non-uniform memory access (NUMA) domain. Therefore, OpenRBC is a powerful tool to investigate biomechanics of the RBC at single cell level. Hybrid continuum-particle-based technique is a cross-fertilization between the continuum and particle-based techniques, and can also facilitate accurate RBC modellings at higher computational efficiency [5, 20, 98]. For example, Lye et al. [98] developed a hybrid continuum-coarse-grained RBC model combining an existing continuum vesicle model with a coarse-grained cytoskeleton, and have simulated the RBC characteristics during cell sedimentation, optical tweezers stretching deformation conditions and motion of a single RBC in a capillary. However, these simulations only consider the stationary shapes of the RBC, and therefore require improvements to consider RBC dynamics. The availability of several RBC models based on variety of numerical techniques facilitate simulation of RBC systems at macroscopic, mesoscopic and microscopic scales; however, one has to carefully select the most suitable RBC model for the investigation to be performed.

The treatment of fluid-RBC and RBC-RBC interactions is another challenge in RBC simulations. For example, DPD and SPH numerical techniques treat the RBC membrane as a set of physical particles and consider the fluid-membrane particle interactions as fluid-fluid interactions. Even though this approach is simple, a critical balance for the number of particles and the simulation time-step is required to achieve numerical consistency as the RBC membrane has negligible mass compared with that of cytoplasm or plasma. Therefore, the adoption of numerical techniques such as IBM provides a more realistic physical approach as it treats the membrane as an immersed boundary [80]. In addition, the microscopic interaction distance between RBC-RBC is generally enlarged in mesoscopic and macroscopic simulations where the RBC-RBC interactions are important due to limitations of computational resources. Therefore, multiscale or hybrid continuum-particle-based numerical techniques are more suitable to investigate RBC behaviour under these circumstances.

1.5 Numerical Predictions of the Equilibrium RBC Morphology

For numerical investigation purposes, the biconcave discocyte shape can be expressed as follows using Cartesian coordinates.

$$z\,(x,y) = \pm D_0 \sqrt{1 - \frac{4\left(x^2+y^2\right)}{D_0^2}} \left[a_0 + a_1 \frac{\left(x^2+y^2\right)}{D_0^2} + a_2 \frac{\left(x^2+y^2\right)^2}{D_0^4} \right],$$

$$-0.5D_0 \le x, y \le 0.5D_0$$

$$(1)$$

where $D_0 = 7.82\ \mu\text{m}$ is the cell diameter, $a_0 = 0.05179025$, $a_1 = 2.002558$ and $a_2 = -4.491048$ [28, 63, 95, 97]. This relationship can produce a biconcave shape having a surface area of $135\ \mu\text{m}^2$ and enclosed volume of $94\ \mu\text{m}^3$, which agrees well with the physiological discocyte shape of the RBC. In addition, the equilibrium RBC shape was derived by minimizing the in-plane stretching energy and the out-of-plane bending energy of the RBC membrane under the reference constraints of cell membrane area and cell volume [13, 20, 99, 100].

The in-plane stretching energy ($E_{Stretching}$) of the RBC membrane is represented by several forms of linear ($E_{Stretching}^{Linear}$) and non-linear (e.g. $E_{Stretching}^{WLC-C}$, $E_{Stretching}^{WLC-POW}$ and $E_{Stretching}^{FENE-C}$) approximations of spring models. $E_{Stretching}^{Linear}$, $E_{Stretching}^{WLC-C}$, $E_{Stretching}^{WLC-POW}$ and $E_{Stretching}^{FENE-C}$ can be given as in Eqs. 2, 3, 4, 5, 6, and 7 [62, 101].

$$E_{Stretching}^{Linear} = \frac{1}{2} k_l \sum_{j=1}^{N_S} \left(l_j - l_{j,0}\right)^2 \qquad (2)$$

where k_l is the linear spring constant, N_S is the number of adjacent vertex-vertex connections of the triangulated membrane surface, l_j is the length of jth link, and $l_{j,0}$ is the equilibrium length of jth link.

$$E_{Stretching}^{WLC-C} = \sum_{j=1}^{N_S} \left[\frac{k_B\,T\,l_{j,\max}}{4\,p} \frac{3x_j^2 - 2x_j^3}{1 - x_j} \right]$$

$$+ \sum_{k=1}^{N_T} \left[\frac{\sqrt{3}\,A_{k,0}^{q+1}\,k_B\,T\,\left(4x_{j,0}^2 - 9x_{j,0} + 6\right)}{4pq\,A_{k,0}^q\,l_{j,\max}\left(1 - x_{j,0}\right)^2} \right] \qquad (3)$$

where k_B is the Boltzmann constant, T is the absolute temperature, $l_{j,\max}$ is the length of jth link at maximum extension, and p is the persistence length. x_j is defined as $x_j = l_j/l_{max}$ and $x_{j,0}$ defined as $x_{j,0} = l_{j,0}/l_{j,\max}$. N_T is the number of triangles and $A_{k,0}$ is the equilibrium area of kth triangle.

$$E_{Stretching}^{WLC-POW} = \sum_{j=1}^{N_S} \left[E_{WLC}\left(l_j\right) + E_{POW}\left(l_j\right) \right] \qquad (4)$$

$$E_{WLC}\left(l_j\right) = \frac{k_B\,T\,l_{j,\max}}{4\,p} \frac{3x_j^2 - 2x_j^3}{1 - x_j} \qquad (5)$$

$$E_{POW}\left(l_j\right) = \begin{cases} \dfrac{k_p}{(m-1)\,l_j^{m-1}}, & m \neq 1 \\ -\,k_p\,\log\left(l_j\right), & m = 1 \end{cases} \tag{6}$$

where k_p is the power function coefficient and m is an exponent such that $m > 0$ [62].

$$E_{Stretching}^{FENE-C} = \sum_{j=1}^{N_S}\left[-\frac{k_s}{2}\,l_{j,\max}^2 \log\left(1 - x_j^2\right)\right] + \sum_{k=1}^{N_T}\left[\frac{\sqrt{3}\,A_0^{q+1}\,k_s}{q\,A_{k,0}^q\left(1 - x_{j,0}^2\right)}\right] \tag{7}$$

where k_s is the FENE spring constant. A brief review of the above in-plane stretching energy models is presented in Table 1.

The linear form is the simplest; however, wormlike chain (WLC) and finitely extensible non-linear elastic (FENE) models give better representation of non-linear nature of spectrin molecules. The maximum extension of the jth link is limited to $l_{j,\,\max}$ as the corresponding spring force reaches infinity when the spring length approaches $l_{j,\,\max}$. The first term of the right hand side of $E_{Stretching}^{WLC-C}$, $E_{Stretching}^{WLC-POW}$

Table 1 Summary of in-plane stretching energy ($E_{Stretching}$) models of the RBC membrane

Model	Eq. #	Main Characteristics	Examples of Applications
$E_{Stretching}^{Linear}$	2	Simplest; less expensive; limited potential to capture the non-linear nature of the spectrin links; individual equilibrium spring lengths can be easily defined	Elastic force of RBC membrane during tank-treading motion [102]; Deformation of RBCs in non-uniform capillaries [101]; RBC membrane mechanics during AFM indentation [88].
$E_{Stretching}^{WLC-C}$	3	Commonly used than the linear, WLC-POW and FENE-C models; better stability at large deformations;	RBC membrane deformations during optical tweezers stretching [103]; RBC membrane mechanics at extreme temperature conditions [104]; RBC rheology in multiscale domains [105].
$E_{Stretching}^{WLC-POW}$	4	First proposed by Fedosov et al. [62]; requires a weak local area constraint for stability at large deformations; individual equilibrium spring lengths can be easily defined	RBC large deformation in a microfluidic system [22]; Stomatocyte-discocyte-echinocyte morphology transformations of a RBC [106]; RBCs in type 2 diabetes mellitus [107].
$E_{Stretching}^{FENE-C}$	7	More rapid spring hardening compared to WLC models at large deformations; less stable at large deformations and requires smaller time steps than WLC;	RBC motion in a capillary [98]; Coarse-graining of spectrin level RBC models [62].

and $E_{Stretching}^{FENE-C}$ represents attractive potentials in the springs, and therefore results in triangular area compression. However, the second term of the right hand side of $E_{Stretching}^{WLC-C}$ and $E_{Stretching}^{FENE-C}$ provides triangular area expansion, and therefore guides the spring length towards equilibrium. $E_{Stretching}^{WLC-POW}$ is not composed of a triangular area expansion term; however, includes a repulsive potential as a power function $(E_{POW}(l_j))$ to restrict the length of the spring. $E_{Stretching}^{WLC-POW}$ was first proposed by Fedosov et al. [62]. In addition, there are continuum models based triangular spring models [108], which can be used to represent the stretching characteristics of the RBC membrane. However, a comprehensive analysis of the suitability of existing stretching models for accurate modelling of RBC mechanical properties, rheology and dynamics, is still lacking to the best of the authors' knowledge.

The out-of-plane bending energy ($E_{Bending}$) of the RBC membrane is represented by the spontaneous curvature model (SCM), bilayer-coupling model (BCM), and area-difference-elasticity (ADE) [13, 109]. The SCM describes the membrane bending energy ($E_{Bending}^{SCM}$) for a membrane with surface area, A, and local bending modulus, κ, such that [109, 110]:

$$E_{Bending}^{SCM}\ [C_1, C_2] = \frac{\kappa}{2} \oint dA\ \left(C_1\ (r) + C_2\ (r) - \overline{C_0}\ \right)^2 \tag{8}$$

where $C_1(r)$ and $C_2(r)$ are the principal curvatures at the point r on the membrane surface, whereas $\overline{C_0}$ is the spontaneous curvature and indicates any asymmetry between the two bilayer-leaflets. Therefore, in the SCM-derived models, vesicle shape is obtained at the minimum of $E_{Bending}^{SCM}$ for given A and vesicle volume, V. The discocyte RBC shape is usually derived from the SCM approach at $\overline{C_0} \sim 0$ [62].

The BCM is based on the bilayer-couple hypothesis and assumes a fixed area for a membrane lipid molecule and no molecular exchange between the two bilayer-leaflets. Therefore, the area of each bilayer-leaflet remains constant, and the area-difference between the two bilayer-leaflets (ΔA) can be determined from the integrated mean curvature over the membrane surface, such that [109]:

$$\Delta A = D \oint dA\ (C_1(r) + C_2(r)) \tag{9}$$

where D is the distance between bilayer-leaflets. The membrane bending energy $E_{Bending}^{BCM}$ in this instance is determined for a defined reference surface with fixed ΔA as another constraint, such that [109]:

$$E_{Bending}^{BCM}\ [C_1, C_2] = \frac{\kappa}{2} \oint dA\ (C_1(r) + C_2(r))^2 \tag{10}$$

The BCM model-derived vesicle shape is obtained at minimum $E_{Bending}^{BCM}$ for given A, V and ΔA. It has been proven elsewhere [100, 110, 111] that both SCM and BCM models lead to the same shape equations, and the vesicle shape behaviour is an extensively studied aspect [18, 21, 100, 109–122]. The ADE model is a combined representation of SCM and BCM, and the ADE model determined membrane energy (E_{ADE}) for a vesicle having A, V and ΔA is as follows [109],

$$E_{ADE}\,[C_1, C_2, \Delta A] = \frac{\kappa}{2}\oint dA\,\left(C_1(r) + C_2(r) - \overline{C_0}\right)^2 + \frac{\overline{\kappa}}{2}\,\frac{\pi}{A\,D^2}\,(\Delta A - \Delta A_0)^2$$

(11)

where $\overline{\kappa}$ is the non-local bending modulus and ΔA_0 is the reference area-difference between bilayer-leaflets. The ADE model converges to the SCM model at $\overline{\kappa}/\kappa \rightarrow 0$, and into BCM at $\overline{\kappa}/\kappa \rightarrow \infty$ [109, 114, 118]. Several studies [100, 109, 110, 118] have comprehensively reviewed these out-of-plane bending energy models for a vesicle. Many research studies have numerically investigated vesicle shapes along with RBC morphologies under a variety of morphology transformation conditions [18, 21, 100, 109, 110, 115, 117–119]. However, the existing numerical predictions of RBC morphology are validated qualitatively only against analogous experimental observations, and a framework of quantitative validation is yet to be implemented. Therefore, the coarse-grained (CG)-RBC membrane model is developed primarily to overcome this limitation. The following sections detail the achievement of the numerical framework such that an improved CG model is developed to accurately predict RBC morphology and deformability.

2 Development of the CG-RBC Membrane Model

Coarse-graining (CG) is a popular particle-based numerical technique, and has several advantages making it the most suitable for the present investigation. A CG-RBC membrane model has better computational efficiency as each membrane particle represents a group of cytoskeletal actin junctional complexes, and therefore, reduces the exhausting number of required particles to discretise the RBC membrane. In addition, the CG technique facilitates the integration of membrane heterogeneity and structural defects, and a greater potential to investigate the influence of storage lesion induced membrane defects on cell morphology and deformability. Furthermore, a CG-RBC membrane model is also suitable to investigate the RBC characteristics under varying other pathophysiological conditions such as hereditary haemolytic disorders (e.g. spherocytosis, elliptocytosis and ovalocytosis) [1, 2], sickle cell disease, malaria [31, 72, 81, 84, 87, 97, 123, 124], and shape-transforming conditions (e.g. stomatocytogenic and echinocytogenic environments) [21, 46, 48, 120, 125–127].

Following subsections describe the formulation of the CG-RBC membrane model for predicting the equilibrium RBC state. Section 2.1 presents the free-energy function of the CG-RBC membrane, and describes the numerical minimization of the overall free-energy of the RBC membrane for given reference conditions such that the equilibrium cell state is achieved. Sections 2.2 and 2.3 describe the construction process of the initial geometry, and the cytoskeletal reference state of the CG-RBC membrane respectively. The computational implementation of the CG-RBC membrane model to achieve the equilibrium RBC state is detailed in Sect. 2.4. Section 2.5 describes the process of CG-RBC membrane model prediction of the healthy discocyte RBC morphology analogous to experimental observations, whereas Sect. 2.6 validates the numerically predicted equilibrium RBC shape against analogous experimental observations.

2.1 Free-Energy of the CG-RBC Membrane

The CG-RBC membrane model is composed of N_V vertices that represent the actin junctional complexes in the RBC membrane cytoskeleton and forms a 2D triangulated surface having N_T triangles. The total free-energy of the CG-RBC membrane (E) is the collective contribution of in-plane stretching energy ($E_{Stretching}$), out-of-plane bending energy ($E_{Bending}$) and the energy penalty to maintain reference cell surface area ($E_{Surface - area}$) and cell volume (E_{Volume}), and is given as [1, 62]:

$$E = E_{Stretching} + E_{Bending} + E_{Surface-area} + E_{Volume} \qquad (12)$$

The N_s adjacent vertex-vertex connections of the triangulated membrane surface represent the spectrin links attached to the actin junctional complexes and contribute to $E_{Stretching}$. $E_{Stretching}$ was estimated based on the coarse-grained $E_{Stretching}^{WLC-POW}$ (Eq. 4) approach implemented by Fedosov et al. [62]. The membrane shear modulus (μ_0) in this instance is given by [62]:

$$\mu_0 = \frac{\sqrt{3}\,k_B\,T}{4\,p\,l_{max}\,x_0}\left[\frac{x_0}{2\,(1-x_0)^3} - \frac{1}{4\,(1-x_0)^2} + \frac{1}{4}\right] + \frac{\sqrt{3}\,k_p\,(m+1)}{4\,l_0^{m+1}} \qquad (13)$$

where l_0 is the equilibrium spectrin link length and defined as $x_0 = l_0/l_{max}$. The parameters k_p and p are estimated for a given μ_0 and x_0 using Eqs. 4 and 13 at the equilibrium cytoskeletal reference state.

$E_{Bending}$ of the RBC membrane was estimated based on the discrete approximation proposed by Jülicher [91] at zero spontaneous membrane curvature, such that:

$$E_{Bending} = 2\,\kappa \sum_{j=1}^{N_S} \frac{M_j^2}{\Delta A_j} \qquad (14)$$

where κ is the membrane bending modulus, M_j is the membrane curvature at the jth link, and ΔA_j is the membrane surface area associated with the jth link. M_j and ΔA_j corresponding to the triangle-pair composed of $T1$ and $T2$ triangles that share the jth link (Fig. 3), were estimated as follows:

$$M_j = \frac{1}{2} l_j \theta_j \qquad (15)$$

$$\Delta A_j = \frac{1}{3} (A_{T1} + A_{T2}) \qquad (16)$$

where θ_j is the angle between outward normal vectors to the triangles $T1$ and $T2$, and A_{T1} and A_{T2} are the planer area associated with $T1$ and $T2$ triangles respectively. θ_j is defined such that the concave arrangement of a triangle-pair corresponds to a positive θ_j, whereas the convex arrangement corresponds to a negative θ_j (Fig. 3), and results in positive or negative M_j respectively.

The energy components of $E_{Surface - area}$ and E_{Volume} were estimated as follows [62, 128]:

$$E_{Surface-area} = \frac{1}{2} k_A \left(\frac{A - A_0}{A_0} \right)^2 A_0 + \sum_{k=1}^{N_T} \frac{1}{2} k_a \left(\frac{A_k - A_{k,0}}{A_{k,0}} \right)^2 A_{k,0} \qquad (17)$$

$$E_{Volume} = \frac{1}{2} k_V \left(\frac{V - V_0}{V_0} \right)^2 V_0 \qquad (18)$$

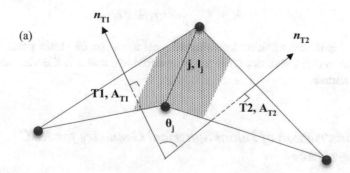

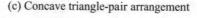

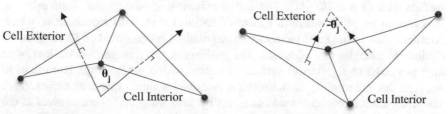

Fig. 3 (a) Illustration of θ_j, l_j, A_{T1} and A_{T2} corresponding to the triangle-pair made of triangles $T1$ and $T2$ triangles that share the jth link, and illustration of (b) convex, and (c) concave triangle-pair arrangements resulting in positive and negative θ_j respectively. n_{T1} and n_{T2} are the normal vectors to the tringles $T1$ and $T2$

where A_0 is the reference membrane surface area, A is the instantaneous membrane surface area, $A_{k,0}$ is the reference area of kth triangle, A_k is the instantaneous area of kth triangle, V_0 is the reference cell volume and V is the instantaneous cell volume. k_A, k_a and k_V represent the total surface area, local surface area and volume constraint coefficients respectively. The resistance of the lipid-bilayer for surface area change was considered for both the whole RBC membrane surface and for the individual triangles, as the lipid-bilayer is anchored to the cytoskeleton through transmembrane proteins, and therefore, the movement of lipid molecules over the membrane is restricted. The first term of the right-hand side of the Eq. 17 represents the total surface area constraint whereas the second term of the right-hand side of the Eq. 17 represents the triangle surface area constraint.

It was assumed that the vertex points move over the RBC membrane surface to achieve the minimum free-energy state, which is the equilibrium RBC shape. The force (F_i) acting on the ith membrane vertex at point r_i on the surface was derived from the principle of virtual work, such that:

$$F_i = -\frac{\partial E}{\partial r_i}, \qquad i \in 1 \dots N_V \tag{19}$$

The resulting motion of the ith membrane vertex was then estimated from the Newton's second law of motion as follows:

$$F_i + f_i^{ext} = m_i \ddot{r}_i + c\,\dot{r}_i \tag{20}$$

where f_i^{ext} is the contribution from any external forces on ith vertex point, m_i is the mass of ith vertex point, dot (.) is the time derivative and c is the viscosity of the RBC membrane.

2.2 Construction of Initial Spherical Geometry for RBC Membrane

The RBC membrane was initially assumed to be a sphere having an equivalent surface area to a RBC (A_0). The initial spherical geometry was built upon an icosahedron inscribed within a sphere of radius 1.0 m. This icosahedron, which constitutes the initial RBC membrane triangulation, is composed of 12 vertices, 20 equilateral triangles and 30 edges. The position of the icosahedron vertices were then projected to a spherical surface of radius with a surface area equivalent to A_0, such that the resulting icosahedron is inscribed within a sphere of radius R_{RBC}. The triangulation refinement was obtained by generating additional vertices at the midpoint of each triangle edge and connecting these new vertices together such that the preceding triangle is divided into four smaller triangles. The new vertices were then projected radially onto the spherical surface with the radius R_{RBC}. The Cartesian coordinates of the new vertex point projected on to the spherical surface are given by:

$$\frac{R_{RBC}}{\sqrt{\left(\frac{x_1+x_2}{2}\right)^2 + \left(\frac{y_1+y_2}{2}\right)^2 + \left(\frac{z_1+z_2}{2}\right)^2}} \left(\frac{x_1+x_2}{2}, \frac{y_1+y_2}{2}, \frac{z_1+z_2}{2}\right) \tag{21}$$

where (x_1, y_1, z_1) and (x_2, y_2, z_2) are the Cartesian coordinates of the icosahedron vertices associated with the edge. The desired level of triangulation was achieved through the successive refinement of the resulting triangulation. The resulting number of membrane triangles (N_T), vertices (N_V), and adjacent vertex-vertex connections (N_S) are given by:

$$N_T = 20 \times \left(4^{N_{Degree}}\right) \tag{22}$$

$$N_V = \frac{1}{2}\ (N_T + 4) \tag{23}$$

$$N_S = N_T + N_S{-}2 \tag{24}$$

where N_{Degree} is the number of triangulation refinement stages. The values of N_T, N_V and N_S at $0 \leq N_{Degree} \leq 6$, are summarised in Table 2.

RBC cytoskeleton has about 27,000–45,000 actin junctional complexes [62], and therefore, the triangulation at $N_{Degree} = 6$ was considered as the spectrin level triangulation. The stages of the first triangulation refinement of the initial icosahedron is presented in Fig. 4.

The subsequent studies presented in this thesis do not consider further remodelling of the triangulated membrane surface, and therefore, N_T, N_V, N_S and their associate interconnections remain constant. Therefore, it is important to identify the refinement of the membrane triangulation such that the triangulation quality and the membrane resolution satisfy the minimum requirements to achieve accurate and efficient numerical predictions. The following subsections detail the assessment of the minimum triangulation quality and the membrane resolution of the CG-RBC membrane model.

Table 2 The values of N_T, N_V and N_S at first six triangulation refinement stages

N_{Degree}	N_T	N_V	N_S
0	20	12	30
1	80	42	120
2	320	162	480
3	1280	642	1920
4	5120	2562	7680
5	20,480	10,242	30,720
6	81,920	40,962	122,880

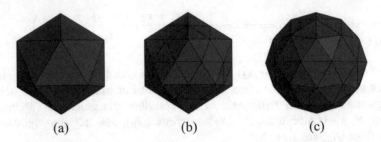

(a)   (b)   (c)

Fig. 4 Stages of first triangulation refinement from initial icosahedron geometry, (**a**) initial icosahedron, (**b**) implementing successive triangulation, and (**c**) projecting the new vertices onto the spherical surface

Table 3 The distribution of edge length ($d(l)$) and vertex attachment to neighbouring triangles ($N_{V\text{-}T}$) for $0 \leq N_{Degree} \leq 6$

N_{Degree}	$d(l)$	Vertices (%) having $N_{V-T} = 6$	Vertices (%) having $N_{V-T} = 5$
0	0.000	0.000	100.000
1	0.104	71.429	28.571
2	0.065	92.593	7.407
3	0.065	98.131	1.869
4	0.065	99.532	0.468
5	0.065	99.883	0.117
6	0.065	99.971	0.029

2.2.1 Required Minimum Triangulation Quality of the RBC Membrane

The triangulation quality was characterized by the distribution of edge length and distribution of the vertex attachment to neighbouring triangles [62]. The distribution of edge length ($d(l)$) was characterised as $d(l) = \sigma(l)/\bar{l}$, where $\sigma(l)$ is the standard deviation for edge length and $\bar{l}$ is the average edge length. The distribution of vertex attachment to neighbouring triangles was characterized by the relative percentage of vertices having a different number of vertex attachment to neighbouring triangles (N_{V-T}). The 12 vertices on the initial icosahedron are each attached to five neighbouring triangles ($N_{V-T} = 5$) whereas additional vertices formed by triangulation refinement are each attached to six neighbouring triangles ($N_{V-T} = 6$). Therefore, the CG-RBC membrane has only two different numbers of vertex attachment to neighbouring triangles. The theoretical estimations of the RBC membrane properties are based on a triangular network composed of equilateral triangles having $N_{V-T} = 6$ [62]. Therefore, higher triangulation quality is achieved at the combination of low $d(l)$ and a higher percentage of vertices having $N_{V-T} = 6$. Table 3 summarises the distribution of $d(l)$ and N_{V-T} for $0 \leq N_{Degree} \leq 6$.

The quality of the triangulation improved with N_{Degree}, and the relative percentage of vertices having $N_{V-T} = 5$ falls below 1.0% for $N_{Degree} \geq 4$. Therefore, the CG-RBC membrane should be at least $N_T \geq 5120$ to achieve reasonable triangulation quality.

2.2.2 Required Minimum Resolution for the CG-RBC Membrane

The required minimum particle resolution of the RBC membrane was determined based on the error between numerically obtained and exact values of the CG-RBC membrane parameters (i.e. $E_{Stretching}$, $E_{Bending}$, A_0 and V_0) of the initial spherical geometry. The exact value of $E_{Stretching}$ for a spherical geometry composed of N_T equilateral triangles was estimated at $\mu_0 = 4.0 \, \mu Nm^{-1}$, $x_0 = 0.45$ and $m = 2$ [62]. The exact value of out-of-plane bending energy ($E^*_{Bending}$) for a closed membrane surface A is given in Eq. 25, and is equivalent to $8 \pi \kappa$ for a spherical surface [122]:

$$E^*_{Bending} = 2 \kappa \oint (2 H)^2 \, dA \equiv 8 \pi \kappa \qquad (25)$$

where H is the spontaneous mean curvature over the surface. Similarly, the exact values of the spherical surface area (A_0^*) and enclosed volume (V_0^*) were estimated as $A_0^* = 4 \pi R_{RBC}^2$ and $V_0^* = 3/4 \pi R_{RBC}^3$. The absolute percentage error (ε) between numerically and theoretically determined values for the spherical geometry for $0 \leq N_{Degree} \leq 6$, is summarised in Table 4, where the subscript of ε denote the corresponding parameter (i.e. $E_{Stretching}$, $E_{Bending}$, A_0 or V_0).

The numerical and theoretical estimations for $E_{Stretching}$, $E_{Bending}$, A_0 and V_0 parameters agreed well for $N_{Degree} \geq 3$, and therefore, the CG-RBC membrane resolution should be at least $N_T \geq 1280$ to achieve reasonable accuracy in numerically predicted results.

Table 4 The absolute percentage error (ε) between numerically and theoretically determined values for the spherical geometry for $0 \leq N_{Degree} \leq 6$. Subscript of ε denote the corresponding parameter in consideration; $E_{Stretching}$, $E_{Bending}$, A_0 and V_0

N_{Degree}	$\varepsilon_{E_{Stretching}}$ (%)	$\varepsilon_{E_{Bending}}$ (%)	ε_{A_0} (%)	ε_{V_0} (%)
0	10.094	23.808	23.808	39.454
1	3.440	6.183	7.166	12.655
2	1.897	0.791	1.882	3.384
3	1.573	0.630	0.477	0.861
4	1.528	0.989	0.120	0.216
5	1.534	1.080	0.030	0.054
6	1.545	1.102	0.008	0.014

The CG-RBC membrane at $N_{Degree} = 4$, which is composed of $N_V = 2562$, $N_T = 5120$ and $N_S = 7680$ was selected as the most suitable triangulation for subsequent studies, since it fulfils the requirements of minimum triangulation quality and membrane resolution.

2.3 Cytoskeletal Reference State of the CG-RBC Membrane

The stress-free cytoskeletal reference state was assumed to be an ellipsoid [115, 117, 122], having a reduced volume of 0.94 of a sphere having $A_{0, Cyto}$ surface area (ν_{Cyto} = 0.94) [106]. A similar approach as in Lim et al. [117] was performed to generate cytoskeletal reference states and the CG-RBC membrane model was adapted to represent only the cytoskeletal spectrin network such that the stable minimum energy state is determined at set reference cytoskeletal surface area ($A_{0, Cyto}$), cytoskeletal volume ($V_{0, Cyto}$) and cytoskeletal reduced volume (ν_{Cyto}). $A_{0, Cyto}$ was assumed to be equivalent to that of the RBC (A_0) whereas the reference triangular element surface area of the cytoskeleton ($A_{k, 0, Cyto}$) was set at the corresponding triangular element area at the initial spherical geometry having the radius R_{RBC}. The presence of $E_{Bending}$, in the form of a stronger bending modulus, weakens the contribution from shear modulus and leads to an unstressed cytoskeletal state. In addition, the presence of cytoskeletal shear modulus, though in weaker form, avoids any numerical inconsistency. Therefore, a significantly higher bending modulus ($\kappa = 5.010^{-18}$Nm) was used at the physiological cytoskeletal shear modulus ($\mu_0 = 4.0$ μNm^{-1}) in order to predict the resultant cytoskeletal equilibrium state. The constraint coefficients; k_A, k_a and k_V were set to 1×10^{-3} Nm^{-1}, 5×10^{-5} Nm^{-1} and 100 Nm^{-2} respectively. The stress-free equilibrium cytoskeletal state was acknowledged at the minimum free-energy state of the triangulated surface at the above cytoskeletal reference conditions, and corresponding l_0 was extracted at ν_{Cyto}. The ellipsoidal stress-free cytoskeletal reference state obtained through the CG-RBC membrane model is presented in Fig. 5.

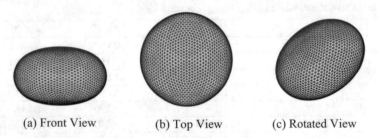

(a) Front View   (b) Top View   (c) Rotated View

Fig. 5 The CG-RBC model predicted cytoskeletal reference state at $\nu_{Cyto} = 0.94$: (a) Front view, (b) Top view, and (c) Rotated view of the cytoskeletal reference state

2.4 Computational Implementation of the CG-RBC Membrane Model

The equilibrium RBC shape was derived under the set reference conditions, where the CG-RBC membrane model particles moved over the space in agreement with Newton's second law of motion and in progressive iterations such that the minimum free-energy state is achieved. The time step (Δt) for succeeding iteration, and m_i and c in Eq. 20 do not affect the equilibrium RBC shape, though control the speed of convergence to the equilibrium state and should be suitably set to avoid any numerical inconsistency. The updated velocity $(\dot{r}_i)$ and the position (r_i) of the ith vertex at the time $(t + \Delta t)$ from time (t) is given as:

$$\dot{r}_i\,(t + \Delta t) = c\,\dot{r}_i(t) + \ddot{r}_i(t)\,\Delta t \tag{26}$$

$$r_i\,(t + \Delta t) = r_i(t) + \dot{r}_i\,(t + \Delta t)\,\Delta t \tag{27}$$

The iterations were continued until the RBC membrane reached the equilibrium state, which is the minimum free-energy state of the RBC membrane at given reference conditions. In the present computational implementation, the equilibrium cell state was acknowledged and the derivation was terminated when the change between each analogous energy component ($E_{Stretching}$, $E_{Bending}$, $E_{Surface\,Area}$, E_{Volume}, $E_{Area-difference}$, $E_{Total-curvature}$) at two successive iterations is less than 1×10^{-7} in the order of energy component in consideration.

The initial spherical geometry was generated according to Sect. 2.2 using MATLAB R2017b, and the source codes for the CG-RBC membrane model was implemented using FORTRAN 90 programming language based on previous work done by Polwaththe-Gallage et al. [101] and Barns et al. [88]. The relationship of the source codes arrangement of the CG-RBC membrane model is presented in Fig. 6. The CG-RBC membrane model computations were carried out on QUT's high performance computing (HPC) resources, and the data analysis and visualization were performed on Microsoft Excel and MATLAB software applications.

2.5 The CG-RBC Membrane Model Predicted Discocyte Morphology

2.5.1 The CG-RBC Membrane Model Parameters for Discocyte Morphology Prediction

The biconcave discocyte morphology was generated initiating from spherical CG-RBC membrane. The physiological RBC surface area being $\sim 140.0\ \mu m^2$ [19, 23], A_0 was selected as $140.0\ \mu m^2$, and therefore, the estimated $R_{RBC} = 3.34\ \mu m$. μ_0 was set at $4.0\ \mu Nm^{-1}$ and agree with the RBC physiological shear modulus [13,

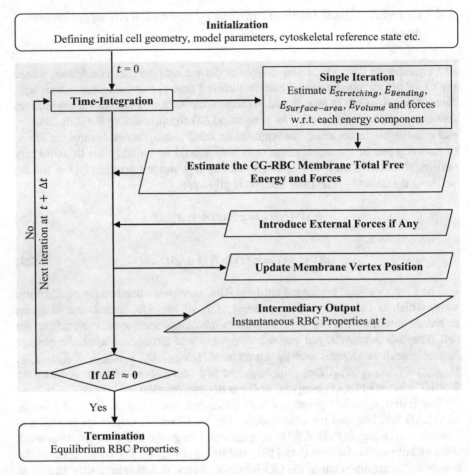

Fig. 6 Relationship of the source codes arrangement in the CG-RBC membrane model

62], and the parameters k_p, p and l_{max} were estimated at $T = 296.15$ K, $m = 2$ and $x_0 = 0.45$ [62]. The experimentally estimated RBC membrane bending modulus lies in the range of $1.0 \times 10^{-19} - 7.0 \times 10^{-19}$ Nm [13, 95], and therefore, κ was selected to be 2.5×10^{-19} Nm. The constraint coefficients; k_A, k_a and k_V were set to 1.0×10^{-3} Nm^{-1}, 5.0×10^{-5} Nm^{-1} and 100.0 Nm^{-2} respectively. $A_{k,0}$ was set at the corresponding triangular element area at cytoskeletal reference state at $v_{Cyto} = 0.94$. V_0 was considered as 93.48 μm^3 in agreement with the physiological RBC volume [19, 23], and is 0.6 of volume of the sphere of radius R_{RBC}. The motion of RBC membrane vertices to reach the equilibrium state was estimated at $c = 1.0 \times 10^{-7}$ Nsm^{-1} and $m_i = 1.0 \times 10^{-9}$ kg.

(a) At $t = 0$ (b) At $t = 0.2$ s   (c) At $t = 0.4$ s   (d) At $t = 0.6$ s

(e) At $t = 0.8$ s   (f) At $t = 1.0$ s   (g) At $t = 1.2$ s   (h) At $t = 2.0$ s

Fig. 7 Evolution of cell shape while reaching the equilibrium discocyte morphology: (**a–h**) represent cell shapes at $t = 0, 0.2, 0.4, 0.6, 0.8, 1.0, 1.2$ and 2.0 s respectively

2.5.2 Evolution of Cell Shape, Membrane Free-Energy and Forces During Discocyte Morphology Prediction

The initial spherical shape progressively reaches the equilibrium discocyte morphology at reference A_0, $A_{k, 0}$ and V_0 constraints. Several instantaneous cell shapes at intermediate time points while the system reaches the equilibrium discocyte morphology are presented in Fig. 7. It can be observed that the CG-RBC membrane gradually reduces its enclosed volume and reaches V_0 at A_0, achieving the biconcave discocyte morphology as the minimum energy state.

The RBC membrane initially has very high free-energy due to the energy penalty from the cell volume constraint as V being much deviated from V_0, and therefore, the forces acting on membrane vertices are considerably high as well. The movement of CG-RBC membrane vertices are initially governed primarily by the energy penalty from the cell volume constraint and later by $E_{Bending}$ energy component while satisfying the reference constraint conditions as well.

2.6 Validation of the CG-RBC Membrane Model Through Predicted Discocyte Morphology

The equilibrium discocyte morphology was compared against experimentally observed discocyte morphology from SEM imaging, and it can be observed (Fig. 8) that these discocyte morphologies qualitatively agree well with each other. These SEM imaging experiments were performed by Dr. M. A. Balanant, a colleague of the 'Red blood cell' research group for her doctoral thesis 'Experimental studies of red blood cells during storage' [129], and the detailed information on SEM experimental protocol is available at https://doi.org/10.17504/protocols.io.yvhfw36.

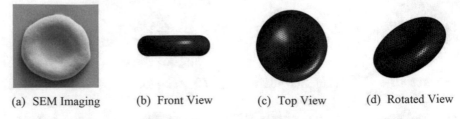

(a) SEM Imaging   (b) Front View   (c) Top View   (d) Rotated View

Fig. 8 Comparison of (**a**) SEM imaging observed (M.A. Balanant, unpublished data, personal communication, March 20, 2018), and (**b**) front view, (**c**) top view, and (**d**) rotated views of the CG-RBC membrane model predicted equilibrium discocyte morphology

In addition to the above qualitative validation, a quantitative analysis was performed on the predicted discocyte morphology against data extracted from 3D confocal microscopy imaging experiments. 3D confocal microscopy imaging data of a randomly selected discocyte RBC morphology were employed to generate an identical triangulated surface mesh, and analogous cell surface area (A^{ex}) and cell volume (V^{ex}) were extracted (Table 5). These 3D confocal microscopy imaging experiments were also performed by Dr. M. A. Balanant for her doctoral thesis [129]; refer to the protocol available at https://doi.org/10.17504/protocols.io.yjyfupw for detailed information of 3D confocal imaging experiments. The equivalent reduced cell volume (v^{ex}) was estimated by:

$$v^{ex} = \frac{V^{ex}}{\text{Volume of a sphere having equivalent } A^{ex}} \qquad (28)$$

Afterwards, the analogous RBC membrane shape was predicted at v^{ex} through the CG-RBC membrane model and resulting cell dimensions were quantitatively compared against the triangulated surface mesh generated from 3D confocal imaging. Assuming rigid body conditions, the centre of mass of the whole cell and its three-principal axes of inertia were determined for each experimentally observed and numerically predicted cell shapes. H_1, H_2 and H_3 are defined as the distance between the furthest vertex points on RBC membrane surface along the three-principal axes of inertia of the cell such that $H_1 \leq H_2 \leq H_3$ (Fig. 9), and used to estimate the cellular measurements: the normalized cell length (H_x), normalized cell thickness (H_z) and shape factor (SF). H_x is defined as the ratio between H_3 and the equivalent spherical radius (R^*), where R^* is the radius of the sphere having an equivalent cell surface area. Similarly, H_z is defined as the ratio between H_1 and R^*. $SF = H_1/\sqrt{H_2 \times H_3}$ [130], and indicates the sphericity of the cell. The cell becomes more spherical as SF reaches the value 1 and becomes a more flattened disc as SF reaches 0.

These three cellular measurements; H_x, H_z and SF were used to quantitatively compare the corresponding experimentally observed and numerically predicted discocyte RBC morphology. The estimated percentage error values (ε) for H_x, H_z

Table 5 Comparison of the CG-RBC membrane model predicted equilibrium discocyte morphology with 3D confocal microscopy imaging observations (M.A. Balanant, unpublished data, personal communication, March 20, 2018). The estimated cell surface area (A^{ex}) and cell volume (V^{ex}) for experimentally observed RBC, normalized cell length (H_x), normalized cell thickness (H_z), and shape factor (SF) are summarised

Confocal Microscopy Imaging	Triangulated Surface from Confocal Microscopy Imaging		A^{ex} (μm^2)	V^{ex} (μm^3)	ν^{ex}
	Top View	Rotated View	145.046	101.623	0.619

CG-RBC Membrane Model Prediction			ε (%)		
Front View	Top View	Rotated View	H_x	H_z	SF
			2.172	20.986	20.995

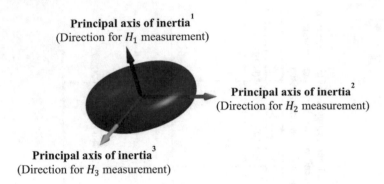

Fig. 9 Measurement of H_1, H_2 and H_3 for a RBC shape. Superscripts 1, 2 and 3 distinguish the three-principal axes of inertia along which H_1, H_2 and H_3 are measured respectively

and SF between corresponding experimentally observed and numerically predicted RBC shapes are presented in Table 5. It can be observed that the equilibrium discocyte morphology obtained through the CG-RBC membrane model agrees well with the morphology derived from 3D confocal microscopy imaging. For instance, the experimentally determined RBC thickness is in the range of 2–2.5 μm whereas RBC diameter is in the range of 6.2–8.2 μm, leading to a maximum percentage deviation of 29% for SF. Therefore, the values of ε for H_x, H_z and SF are reasonable and the maximum ε (= 20.995%) is for the SF. The discocyte shape observed under confocal microscopy imaging not being a completely flattened cell, leads to a higher H_z value, which also affects SF determination. Therefore, ε for H_z and SF between experimentally observed and the CG-RBC model predicted discocyte shape indicates higher values. Furthermore, any experiment error during 3D confocal microscopy imaging and image analysis can contribute to ε while triangulated surface generation can also be another contributing factor. Therefore, the resulting ε is the resultant effect of any experimental errors, any triangulated surface generating errors and any error in the CG-RBC shape predictions. The maximum ε being 20.995% is acceptable based on all these uncertainties, and therefore, the CG-RBC model is capable of quantitatively representing equilibrium RBC shape as well. The present quantitative comparison considered the cellular measurements corresponding to a single cell only. However, the availability of experimentally extracted cellular information on multiple discocyte cells can provide average cellular information for better comparison between experimentally observed and numerically predicted RBC morphology.

This section detailed the development of the CG-RBC membrane model for predicting the equilibrium RBC shape for given reference conditions. Then the deformation behaviour of the CG-RBC membrane model predicted discocyte morphology is investigated in the following section under optical tweezers stretching conditions.

3 Deformation Behaviour of Discocyte During Optical Tweezers Stretching

Cell deformability is a potential criterion to examine the health of a RBC subjected to morphological, structural and functional changes. Deformability investigations provide valuable insights into the physiology, cell biology and biorheology under such pathophysiological conditions. Different techniques have been used to investigate the RBC deformability in vitro, and descriptions of these techniques can be found in [4, 23, 30, 31, 62, 63, 72, 74, 75, 79, 84, 88, 94, 95, 101]. Optical tweezers is one such technique and provides a highly sensitive assessment of the cell deformability at the single-cell level. With optical tweezers, it is possible to trap, manipulate and displace a living cell or a part of it without damage, either directly or using specific handles such as dielectric beads of silica [2, 87, 131–133]. The RBC membrane is primarily responsible for the cell morphology and its elastic response during optical tweezers stretching deformation as the intracellular fluid is purely viscous and has no elasticity [131]. Therefore, the CG-RBC membrane model was employed to investigate the deformability characteristics of a discocyte during optical tweezers stretching deformation. Section 3.1 details the numerical implementation of optical tweezers stretching forces on the RBC, whereas Sect. 3.2 discusses the validation of the numerically predicted discocyte deformation behaviour against reported experimental observation.

3.1 Implementation of Optical Tweezers Stretching

The impact of optical tweezers stretching forces on the CG-RBC membrane model predicted discocyte RBC morphology were numerically investigated. Assuming rigid body conditions, the centre of mass of the whole cell and its three-principal axes of inertia were determined for each RBC morphology. The total stretching force (F^{ext}) was applied on $N_+ = a\, N_V$ vertices whereas $-F^{ext}$ is applied on $N_- = a\, N_V$ vertices along the principal axis of inertia[3] (Fig. 10) [1, 62]. N_+ and N_- are the vertices that locate within the circular region of radius $d_C/2$ on the initial spherical geometry, and from the two vertices ($i_{X_{max}}$ and $i_{X_{min}}$) on the furthest ends of the equilibrium cell shape along the principal axis of inertia[3] (Fig. 10). d_C is the contact diameter between the cell membrane and the attached silica beads. The vertex fraction a corresponds to d_C and is given by $a = \pi\, d_C^2 / (4\, A_0)$. Therefore, f_i^{ext} was applied on ith vertex, such that:

$$f_i^{ext} = \begin{cases} F^{ext}/N_+, & i \in N_+ \\ -F^{ext}/N_-, & i \in N_- \\ 0, & i \notin (N_+ \cup N_-) \end{cases} \tag{29}$$

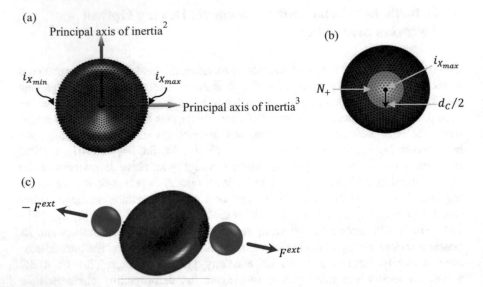

Fig. 10 Optical tweezers stretching implementation on discocyte cell: (**a**) identification of $i_{X_{max}}$ and $i_{X_{min}}$ vertices, (**b**) identification of N_+ vertices, and (**c**) contact region between RBC membrane and binding silica beads

F^{ext} was gradually applied on the cell via regular force increments of ΔF^{ext} where the cell was provided with sufficient time to converge to the equilibrium stretched state after each force increment. The equilibrium stretched cell state was determined at equivalent A_0, $A_{k,0}$, and V_0 of discocyte at the corresponding F^{ext}. The resultant force and motion of *ith* membrane vertex were determined according to the principle of virtual work (Eq. 19) and Newton's second law of motion (Eq. 20) respectively. Equivalently, the computational implementation considered F^{ext} as the applied external force during time integration (Fig. 6), and the equilibrium cell state was determined.

3.2 Validation of the CG-RBC Membrane Model Predicted Discocyte Deformation Behaviour

The optical tweezers stretching deformation of the equilibrium discocyte cell was investigated at equivalent reference conditions and model parameters as in Sect. 2.5. Analogous to experimental optical tweezers stretching experiments by Suresh et al. [64], F^{ext} was applied on the cell such that $0 \leq F^{ext} \leq 200.0$ pN. d_C was approximated as 2.0 μm [1, 2, 62, 64, 123], and accordingly F^{ext} was applied on $a = 0.02$ membrane vertices in regular increments of $\Delta F^{ext} = 10.0$ pN. The evolution of axial diameter (D_A) (measured along the principal axis of inertia3) and transverse diameter (D_T) (measured along the principal axis of inertia2) of the equilibrium

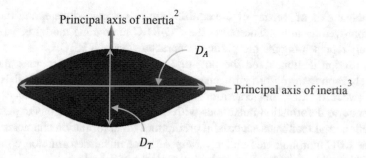

Fig. 11 Measurement of axial (D_A) and transverse (D_T) diameters of a RBC at equilibrium cell stretched state

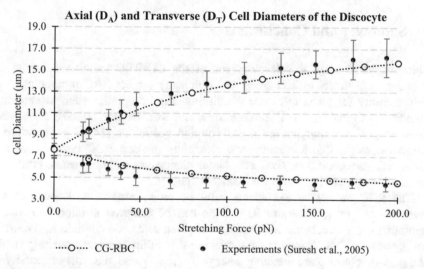

····○···· CG-RBC   • Experiements (Suresh et al., 2005)

Fig. 12 Comparison between the CG-RBC membrane model predicted deformation behaviour at $k_{AD} = 7.5 \times 10^{-17}$ Nm and $k_{AD} = 7.5 \times 10^{-18}$ Nm, versus experimentally observed [64] deformation behaviour of discocyte cell undergoing optical tweezers stretching. The curves in the top represent the evolution of axial diameter (D_A) whereas the curves in the bottom represent the evolution of transverse diameter (D_T)

stretched cell state at F^{ext}, were then compared against experimental observations by Suresh et al. [64]. Refer to Fig. 11 for a graphical representation of D_A and D_T.

It was observed (Fig. 12) that the deformation behaviour of the discocyte agrees very well with the analogous experimental observations. The maximum deviation (ε_{OTS}) between numerically predicted and experimentally observed D_A is only 4.78%. The disagreement in the D_T may be partially due to experimental errors arising from the fact that the optical tweezers stretching measurements have been performed from a single observation angle [62, 63]. A RBC undergoing stretching may rotate on the plane perpendicular to the initial principal axis of inertia[3], and therefore, measurements from a single observation angle may lead to deviations from actual D_T. The numerically estimated D_T was measured along the principal axis of inertia[2] of the equilibrium cell stretched state, and therefore, gives improved accuracy. Therefore, the CG-RBC membrane model is capable of accurately representing the morphology characteristics of a RBC.

This section demonstrated the potential of the CG-RBC membrane model to accurately capture the deformation characteristics of a healthy discocyte RBC under optical tweezers stretching conditions. The numerical approach produces agreeable discocyte deformation behaviour with comparison to analogous experimental observations and facilitates successful predictions on deformation characteristics of different RBC morphologies under variety of mechanical deformation conditions. Following section presents the concluding remarks of this chapter.

4 Summary and Conclusions

This chapter detailed the development of the CG-RBC membrane model; an improved CG model to accurately and efficiently predict RBC morphology and deformability for given reference conditions. RBC properties such as cell morphology, deformability, cell surface area and volume are affected by several diseased [13] and morphology transformation conditions [14, 52]. Therefore, a numerical model that accurately and efficiently predicts RBC characteristics at given reference conditions facilitates better diagnostics and treatments. Following concluding remarks can be drawn from this chapter.

The RBC membrane can successfully be represented as a network of CG particles, and the equilibrium RBC state can be achieved through numerically minimizing the membrane free-energy for given reference conditions. The free-energy of the CG-RBC membrane model consists of the in-plane stretching energy ($E_{Stretching}$), out-of-plane bending energy ($E_{Bending}$) and the energy penalty to maintain reference cell surface area ($E_{Surface - area}$) and cell volume (E_{Volume}). The numerical implementation successfully predicts the equilibrium state of a healthy biconcave discocyte cell, which qualitatively and quantitatively agrees well with analogous experimental observations. In addition to the usual qualitative validation, a quantitative analysis was performed to improve the accuracy of the numerically predicted equilibrium RBC state against analogous experimental observations. In addition, the numerical approach produces agreeable discocyte deformation behaviour with comparison to analogous experimental observations. Therefore, the CG-RBC membrane model represents an improved RBC membrane model to accurately and efficiently predict the discocyte morphology and deformability of a healthy RBC.

This study facilitates comprehensive knowledge of morphological and deformability characteristics of RBCs under several shape-transforming conditions. Geekiyanage et al. [106] discuss the application of the CG-RBC membrane model to accurately represent the complete sequence of RBC morphologies and their deformability associated with the echinocytogenic and stomatocytogenic shape-transforming environments. However, there are some limitations associated with the CG-RBC membrane model that need to be clarified. The quantitative validation of the CG-RBC model predicted discocyte morphology is performed against experimental observations of a single RBC. However, the availability of cellular information on multiple RBCs can provide average cellular information for more accurate numerical predictions. In addition, there is insufficient information on cytoskeletal reference state. The stress-free cytoskeletal reference state is a controversial subject, and this study assumes an ellipsoidal cytoskeletal reference state having a reduced volume of 0.94 of a sphere with equivalent cell surface area, which has been validated in Geekiyanage et al. [106]. However, these investigations were based on the assumptions that the cytoskeletal surface area is equivalent to that of the RBC and is at no pre-stressed condition. It was observed that the numerically predicted morphological and deformability characteristics are affected by the choice of cytoskeletal reference state [103, 106, 117], and therefore, improved understanding on the exact cytoskeletal reference state is required. The CG-RBC membrane model represents the composite RBC membrane and does not explicitly model the cytoplasmic or extracellular fluid components. The dynamics of the RBC while reaching the equilibrium cell state at given reference conditions and during mechanical deformations are influenced by the cytoplasmic and environmental factors. Therefore, the presence of fluid components is required to accurately capture the associated contribution on RBC morphological and deformability characteristics.

However, the CG-RBC membrane model is an improved and general numerical approach to investigate the biomechanics of the RBC membrane with respect to its morphological and deformability changes. Several future research directions that can be built on the findings of the present study are as follows. The present numerical framework can be applied to investigate the deformation characteristics of RBCs having different morphologies subjected to varied mechanical deformation scenarios (e.g. during the passage through microfluidics and the in vivo microcirculation, micropipette aspiration, and AFM indentation). These studies can facilitate improved understanding on RBC biomechanics with respect to its morphology under varying loading configurations. The CG-RBC membrane model can be easily refined to represent the detailed RBC membrane structure including its heterogeneity and any structural defects, and interactions between the bilayer-leaflets and the cytoskeleton for improved morphology predictions that are associated with the cell age, diseased conditions, and extracellular environmental conditions. Introduction of the membrane structural remodelling can facilitate the successful predictions of the RBC membrane vesiculation observed at the higher strength of above shape-transforming conditions, splenic sequestration of less deformable RBCs, and several hereditary haemolytic disorders (e.g. spherocytosis, elliptocytosis and ovalocytosis) for better diagnostics and treatments. Different haemolytic disorders affect the RBC geometry and cellular properties distinctively. For example, the cytoskeleton detaches from the lipid bilayer due to the defects in ankyrin, protein 4.2 or band-3 proteins in HS [13]. The cell biomechanical properties are affected by the Plasmodium parasite in malaria infected RBCs [13]. In addition, sickle-cell anaemia is caused by the intracellular HbS polymerization due to a single point mutation in haemoglobin causing RBC sickling [13]. Therefore, pre-diagnostics can help identify the most sensitive CG-RBC model parameters for a comprehensive study of RBC mechanics and hemodynamic characteristics under these pathophysiological conditions for better diagnostics and treatments. For example, the cytoskeletal defects under HE diseased condition can be incorporated into the CG-RBC membrane model by setting the equilibrium length of the spectrin links individually, whereas the disruption of vertical connections between lipid-bilayer and cytoskeletal actin junctions is incorporated into the model by cancelling the in-plane shear deformation for the spectrin links attached to these actin junctions. The loss of membrane surface in HS is possible to be introduced to the model by appropriate adjustments to reduced cell volume. However, the cytoskeleton is under compression as it is attached to a lipid-bilayer having a lower surface area than a healthy cell. Therefore, the cytoskeletal reference state would need adjustments to discuss HS cell behaviour through the CG-RBC membrane model. Therefore, this study facilitates many investigations into RBC morphology and deformability under diverse shape-transforming scenarios, in vitro RBC storage, microvascular circulation and flow through microfluidic devices.

Bibliography

1. M. Dao, J. Li, S. Suresh, Molecularly based analysis of deformation of spectrin network and human erythrocyte. Mater. Sci. Eng. C **26**(8), 1232–1244 (2006)
2. M. Dao, C.T. Lim, S. Suresh, Mechanics of the human red blood cell deformed by optical tweezers. J. Mech. Phys. Solids **51**(11), 2259–2280 (2003)
3. D. Kuzman, S. Svetina, R.E. Waugh, B. Žekš, Elastic properties of the red blood cell membrane that determine echinocyte deformability. Eur. Biophys. J. **33**(1) (2004)
4. Y. Kim, K. Kim, Y.K. Park, Measurement techniques for red blood cell deformability: Recent advances. INTECH, 167–194 (2012)
5. H.-Y. Chang, X. Li, H. Li, G.E. Karniadakis, MD/DPD multiscale framework for predicting morphology and stresses of red blood cells in health and disease. PLoS Comput. Biol. **12**(10) (2016). Art. no. 1005173
6. X. Li, M. Dao, G. Lykotrafitis, G.E. Karniadakis, Biomechanics and biorheology of red blood cells in sickle cell anemia. J. Biomech. **50**, 34–41 (2016)
7. H. Li, L. Lu, X. Li, P.A. Buffet, M. Dao, G.E. Karniadakis, S. Suresh, Mechanics of diseased red blood cells in human spleen and consequences for hereditary blood disorders. Proc. Natl. Acad. Sci. U. S. A. **115**(38), 9574–9579 (2018)
8. Y. Wang, G. You, P. Chen, J. Li, G. Chen, B. Wang, P. Li, D. Han, H. Zhou, L. Zhao, The mechanical properties of stored red blood cells measured by a convenient microfluidic approach combining with mathematic model. Biomicrofluidics **10**(2) (2016). Art. no. 024104
9. E. Kozlova, A. Chernysh, V. Moroz, V. Sergunova, O. Gudkova, E. Manchenko, Morphology, membrane nanostructure and stiffness for quality assessment of packed red blood cells. Sci. Rep. **7**(1) (2017). Art. no. 7846
10. K. Matthews, M.-E. Myrand-Lapierre, R.R. Ang, S.P. Duffy, M.D. Scott, H. Ma, Microfluidic deformability analysis of the red cell storage lesion. J. Biomech. **48**(15), 4065–4072 (2015)
11. H. Song, Y. Liu, B. Zhang, K. Tian, P. Zhu, H. Lu, Q. Tang, Study of in vitro RBCs membrane elasticity with AOD scanning optical tweezers. Biomed. Opt. Express **8**(1), 384–394 (2017)

12. Y. Zheng, J. Chen, T. Cui, N. Shehata, C. Wang, Y. Sun, Characterization of red blood cell deformability change during blood storage. Lab Chip **14**(3), 577–583 (2014)
13. X. Li, H. Li, H.-Y. Chang, G. Lykotrafitis, G.E. Karniadakis, Computational biomechanics of human red blood cells in hematological disorders. J. Biomech. Eng. **139**(2) (2017). Art. no. 021008
14. P. Wong, A basis of echinocytosis and stomatocytosis in the disc–sphere transformations of the erythrocyte. J. Theor. Biol. **196**(3), 343–361 (1999)
15. Y. Li, C. Wen, H. Xie, A. Ye, Y. Yin, Mechanical property analysis of stored red blood cell using optical tweezers. Colloids Surf. B: Biointerfaces **70**(2), 169–173 (2009)
16. E. Pretorius, The adaptability of red blood cells. Cardiovasc. Diabetol. **12** (2013). Art. no. 63
17. A.V. Buys, M.-J.V. Rooy, P. Soma, D.V. Papendorp, B. Lipinski, E. Pretorius, Changes in red blood cell membrane structure in type 2 diabetes: A scanning electron and atomic force microscopy study. Cardiovasc. Diabetol. **12** (2013). Art. no. 25
18. G.H.W. Lim, M. Wortis, R. Mukhopadhyay, Stomatocyte–discocyte–echinocyte sequence of the human red blood cell: Evidence for the bilayer– Couple hypothesis from membrane mechanics. Proc. Natl. Acad. Sci. U. S. A. **99**(26), 16766–16769 (2002)
19. N. Mohandas, P.G. Gallagher, Red cell membrane: Past, present, and future. Blood **112**(10), 3939–3948 (2008)
20. X. Li, P.M. Vlahovska, G.E. Karniadakis, Continuum- and particle-based modeling of shapes and dynamics of red blood cells in health and disease. Soft Matter **9**(1), 28–37 (2013)
21. R. Mukhopadhyay, G. Lim, M. Wortis, Echinocyte shapes: Bending, stretching, and shear determine spicule shape and spacing. Biophys. J. **82**(4), 1756–1772 (2002)
22. X. Li, Z. Peng, H. Lei, M. Dao, G.E. Karniadakis, Probing red blood cell mechanics, rheology and dynamics with a two-component multi-scale model. Philos. Trans. Royal Soc. A Math. Phys. Eng. Sci. **372** (2014). Art. no. 20130389
23. G. Tomaiuolo, Biomechanical properties of red blood cells in health and disease towards microfluidics, in *Biomicrofluidics*. vol 8(5), (2014), Art. no. 051501
24. J. Li, G. Lykotrafitis, M. Dao, S. Suresh, Cytoskeletal dynamics of human erythrocyte. Proc. Natl. Acad. Sci. U. S. A. **104**(12), 4937–4942 (2007)
25. T. Auth, S.A. Safran, N.S. Gov, Fluctuations of coupled fluid and solid membranes with application to red blood cells. Phys. Rev. E **76**(5) (2007). Art. no. 051910
26. M.P. Sheetz, S.J. Singer, Biological membranes as bilayer couples: Molecular mechanisms of drug-erythrocyte interactions. Proc. Natl. Acad. Sci. U. S. A. **71**(11), 4457–4461 (1974)
27. G. Pages, T.W. Yau, P.W. Kuchel, Erythrocyte shape reversion from echinocytes to discocytes: Kinetics via fast-measurement NMR diffusion-diffraction. Magn. Reson. Med. **64**(3), 645–652 (2010)
28. Z. Peng, R.J. Asaro, Q. Zhu, Multiscale simulation of erythrocyte membranes. Phys. Rev. E **81**(3) (2010). Art. no. 031904
29. H. Li, G. Lykotrafitis, Two-component coarse-grained molecular-dynamics model for the human erythrocyte membrane. Biophys. J. **102**(1), 75–84 (2012)
30. Z. Peng, R.J. Asaro, Q. Zhu, Multiscale modelling of erythrocytes in stokes flow. J. Fluid Mech. **686**, 299–337 (2011)
31. D.A. Fedosov, M. Dao, G.E. Karniadakis, S. Suresh, Computational biorheology of human blood flow in health and disease. Ann. Biomed. Eng. **42**(2), 368–387 (2014)
32. H. Li, G. Lykotrafitis, Erythrocyte membrane model with explicit description of the lipid bilayer and the spectrin network. Biophys. J. **107**(3), 642–653 (2014)
33. H. Li, G. Lykotrafitis, Vesiculation of healthy and defective red blood cells. Phys. Rev. E **92**(1) (2015). Art. no. 012715
34. S. Salehyar, Q. Zhu, Effects of stiffness and volume on the transit time of an erythrocyte through a slit. Biomech. Model. Mechanobiol. **16**(3), 921–931 (2016)
35. M.M. Gedde, D.K. Davis, W.H. Huestis, Cytoplasmic pH and human erythrocyte shape. Biophys. J. **72**(3), 1234–1246 (1997)
36. R. Glaser, The shape of red blood cells as a function of membrane potential and temperature. J. Membr. Biol. **51**, 217–228 (1979)

37. M. Gros, S. Vrhovec, M. Brumen, S. Svetina, B. Zeks, Low pH induced shape changes and vesiculation of human erythrocytes. Gen. Physiol. Biophys. **15**(2), 145–163 (1996)
38. F. Xing, S. Xun, Y. Zhu, F. Hu, I. Drevenšek-Olenik, X. Zhang, L. Pan, J. Xu, Microfluidic assemblies designed for assessment of drug effects on deformability of human erythrocytes. Biochem. Biophys. Res. Commun. **512**(2), 303–309 (2019)
39. T. Avsievich, A. Popov, A. Bykov, I. Meglinski, Mutual interaction of red blood cells influenced by nanoparticles. Sci. Rep. **9**(1) (2019). Art. no. 5147
40. Y.J. Choi, H. Huh, G.E. Bae, E.J. Ko, S.-u. Choi, S.-H. Park, C.H. Lim, H.W. Shin, H.-w. Lee, S.Z. Yoon, Effect of varying external pneumatic pressure on hemolysis and red blood cell elongation index in fresh and aged blood: Randomized laboratory research. Medicine **97**(28) (2018). Art. no. 11460
41. I.I. Jeican, H. Matei, A. Istrate, E. Mironescu, S. Balici, Changes observed in erythrocyte cells exposed to an alternating current. Clujul Medical **90**(2), 154–160 (2017)
42. M.M. Gedde, W.H. Huestis, Membrane potential and human erythrocyte shape. Biophys. J. **72**(3), 1220–1233 (1997)
43. M.M. Gedde, E. Yang, W.H. Huestis, Shape response of human erythrocytes to altered cell pH. Blood **86**(4), 1595–1599 (1995)
44. M.M. Gedde, E. Yang, W.H. Huestis, Resolution of the paradox of red cell shape changes in low and high pH. Biochim. Biophys. Acta Biomembr. **1417**(2), 246–253 (1999)
45. K.D. Tachev, K.D. Danov, P.A. Kralchevsky, On the mechanism of stomatocyte-echinocyte transformations of red blood cells: Experiment and theoretical model. Colloids Surf. B: Biointerfaces **34**(2), 123–140 (2004)
46. S.V. Rudenko, M.K. Saeid, Reconstruction of erythrocyte shape during modified morphological response. Biochem. Mosc. **75**(8), 1025–1031 (2010)
47. S.V. Rudenko, Characterization of morphological response of red cells in a sucrose solution. Blood Cell Mol. Dis. **42**(3), 252–261 (2009)
48. S.V. Rudenko, Erythrocyte morphological states, phases, transitions and trajectories. Biochim. Biophys. Acta Biomembr. **1798**(9), 1767–1778 (2010)
49. S.V. Rudenko, Low concentration of extracellular hemoglobin affects shape of RBC in low ion strength sucrose solution. Bioelectrochemistry **75**(1), 19–25 (2009)
50. B. Deuticke, Transformation and restoration of biconcave shape of human erythrocytes induced by amphiphilic agents and changes of ionic environment. Biochim. Biophys. Acta Biomembr. **163**(4), 494–500 (1968)
51. M. Rasia, A. Bollini, Red blood cell shape as a function of medium's ionic strength and pH. Biochim. Biophys. Acta Biomembr. **1372**(2), 198–204 (1998)
52. M. Bessis, Red cell shapes. An illustrated classification and its rationale, in *Red Cell Shape*. vol 12(6), (Springer, Berlin/Heidelberg, 1973), pp. 721–746
53. G. Brecher, M. Bessis, Present status of spiculed red cells and their relationship to the discocyte-echinocyte transformation: A critical review. Blood **40**(3), 333–344 (1972)
54. I. Mustafa, A. Al Marwani, K.M. Nasr, N.A. Kano, T. Hadwan, Time dependent assessment of morphological changes: Leukodepleted packed red blood cells stored in SAGM. Biomed Res. Int. (2016). Art. no. 4529434
55. R.E. Waugh, M. Narla, C.W. Jackson, T.J. Mueller, T. Suzuki, G.L. Dale, Rheologic properties of senescent erythrocytes: Loss of surface area and volume with red blood cell age. Blood **79**(5), 1351–1358 (1992)
56. D. Kuzman, T. Žnidarčič, M. Gros, S. Vrhovec, S. Svetina, B. Žekš, Effect of pH on red blood cell deformability. Eur. J. Phys. **440**(1), 193–194 (2000)
57. K. Jaferzadeh, I. Moon, Quantitative investigation of red blood cell three-dimensional geometric and chemical changes in the storage lesion using digital holographic microscopy. J. Biomed. Optics **20**(11) (2015). Art. no. 111218
58. S. Piomelli, C. Seaman, Mechanism of red blood cell aging: Relationship of cell density and cell age. Am. J. Hematol. **42**(1), 46–52 (1993)
59. D. Yoon, D. You, Continuum modeling of deformation and aggregation of red blood cells. J. Biomech. **49**(11), 2267–2279 (2016)

60. S.K. Boey, D.H. Boal, D.E. Discher, Simulations of the erythrocyte cytoskeleton at large deformation: I. microscopic models. Biophys. J. **75**(3), 1573–1583 (1998)
61. D.H. Boal, Computer simulation of a model network for the erythrocyte cytoskeleton. Biophys. J. **67**(2), 521–529 (1994)
62. D.A. Fedosov, B. Caswell, G.E. Karniadakis, Systematic coarse-graining of spectrin-level red blood cell models. Comput. Methods Appl. Mech. Eng. **199**(29–32), 1937–1948 (2010)
63. D.A. Fedosov, B. Caswell, G.E. Karniadakis, A multiscale red blood cell model with accurate mechanics, rheology, and dynamics. Biophys. J. **98**(10), 2215–2225 (2010)
64. S. Suresh, J. Spatz, J.P. Mills, A. Micoulet, M. Dao, C.T. Lim, M. Beil, T. Seufferlein, Connections between single-cell biomechanics and human disease states: Gastrointestinal cancer and malaria. Acta Biomater. **1**(1), 15–30 (2005)
65. W. Groner, N. Mohandas, M. Bessis, New optical technique for measuring erythrocyte deformability with the ektacytometer. Clin. Chem. **26**(10), 1435–1442 (1980)
66. X.Y. Chen, Y.X. Huang, W.J. Liu, Z.J. Yuan, Membrane surface charge and morphological and mechanical properties of young and old erythrocytes. Curr. Appl. Phys. **7**, 94–96 (2007)
67. R.T. Card, N. Mohandas, P.L. Mollison, Relationship of post-transfusion viability to deformability of stored red-cells. Br. J. Haematol. **53**(2), 237–240 (1983)
68. L. Van De Watering, More data on red blood cell storage could clarify confusing clinical outcomes. Transfusion **54**(3), 501–502 (2014)
69. R.R. Huruta, M.L. Barjas-Castro, S.T.O. Saad, F.F. Costa, A. Fontes, L.C. Barbosa, C.L. Cesar, Mechanical properties of stored red blood cells using optical tweezers. Blood **92**(8), 2975–2977 (1998)
70. B. Bhaduri, M. Kandel, C. Brugnara, K. Tangella, G. Popescu, Optical assay of erythrocyte function in banked blood. Sci. Rep. **4** (2014). Art. no. 6211
71. S. Ramirez-Arcos, D.C. Marks, J.P. Acker, W.P. Sheffield, Quality and safety of blood products. J. Blood Transfus. **2016**, 1–2 (2016)
72. Y. Park, C.A. Best, K. Badizadegan, R.R. Dasari, M.S. Feld, T. Kuriabova, M.L. Henle, A.J. Levine, G. Popescu, Measurement of red blood cell mechanics during morphological changes. Proc. Natl. Acad. Sci. U. S. A. **107**(15), 6731–6736 (2010)
73. D.C. Betticher, W.H. Reinhart, J. Geiser, Effect of RBC shape and deformability on pulmonary O2 diffusing capacity and resistance to flow in rabbit lungs. J. Appl. Physiol. **78**(3), 778–783 (1995)
74. C. Monzel, K. Sengupta, Measuring shape fluctuations in biological membranes. J. Phys. D Appl.Phys. **49**(24) (2016). Art. no. 243002
75. D. Bento, R. Rodrigues, V. Faustino, D. Pinho, C. Fernandes, A. Pereira, V. Garcia, J. Miranda, R. Lima, Deformation of red blood cells, air bubbles, and droplets in microfluidic devices: Flow visualizations and measurements. Micromachines **9**(4), 3–18 (2018). Art. no. 151
76. M. Musielak, Red blood cell-deformability measurement: Review of techniques. Clin. Hemorheol. Microcirc. **42**(1), 47–64 (2009)
77. X. Li, H. Lu, Z. Peng, Continuum- and particle-based modeling of human red blood cells, in *Handbook of Materials Modeling: Applications: Current and Emerging Materials*, ed. by W. Andreoni, S. Yip, (Springer International Publishing, 2018)
78. A.K. Dasanna, U.S. Schwarz, G. Gompper, D.A. Fedosov, Multiscale modeling of malaria-infected red blood cells, in *Handbook of Materials Modeling: Applications: Current and Emerging Materials*, ed. by W. Andreoni, S. Yip, (Springer International Publishing, 2018)
79. M. Ju, S.S. Ye, B. Namgung, S. Cho, H.T. Low, H.L. Leo, S. Kim, A review of numerical methods for red blood cell flow simulation. Comput. Methods Biomech. Biomed. Engin. **18**(2), 130–140 (2015)
80. T. Ye, N. Phan-Thien, C.T. Lim, Particle-based simulations of red blood cells - a review. J. Biomech. **49**(11), 2255–2266 (2016)
81. A. Yazdani, X. Li, G.E. Karniadakis, Dynamic and rheological properties of soft biological cell suspensions. Rheol. Acta **55**(6), 433–449 (2016)

82. Y. Imai, T. Omori, Y. Shimogonya, T. Yamaguchi, T. Ishikawa, Numerical methods for simulating blood flow at macro, micro, and multi scales. J. Biomech. **49**(11), 2221–2228 (2016)
83. G. Gompper, D.A. Fedosov, Modeling microcirculatory blood flow: Current state and future perspectives. Wiley Interdiscip. Rev. Syst. Biol. Med. **8**(2), 157–168 (2016)
84. D.A. Fedosov, H. Noguchi, G. Gompper, Multiscale modeling of blood flow: From single cells to blood rheology. Biomech. Model. Mechanobiol. **13**(2), 239–258 (2014)
85. J.B. Freund, in *Annual Review of Fluid Mechanics*, ed. by S. H. Davis, P. Moin, Numerical simulation of flowing blood cells, vol 46 (Annual Review of Fluid Mechanics, 2014), pp. 67–95
86. R.C.H. Van Der Burgt, *A Cross-Slot Microrheometer to Probe Red Blood Cell Dynamics* (Technische Universiteit Eindhoven, 2016)
87. C.T. Lim, M. Dao, S. Suresh, C.H. Sow, K.T. Chew, Large deformation of living cells using laser traps. Acta Mater. **52**(7), 1837–1845 (2004)
88. S. Barns, M.A. Balanant, E. Sauret, R. Flower, S. Saha, Y. Gu, Investigation of red blood cell mechanical properties using AFM indentation and coarse-grained particle method. Biomed. Eng. Online **16**(1) (2017). Art. no. 140
89. G. Marcelli, K.H. Parker, C.P. Winlove, Thermal fluctuations of red blood cell membrane via a constant-area particle-dynamics model. Biophys. J. **89**(4), 2473–2480 (2005)
90. I.V. Pivkin, G.E. Karniadakis, Accurate coarse-grained modeling of red blood cells. Phys. Rev. Lett. **101**(11) (2008). Art. no. 118105
91. K.-i. Tsubota, Short note on the bending models for a membrane in capsule mechanics: Comparison between continuum and discrete models. J. Comput. Phys. **277**, 320–328 (2014)
92. Y.-H. Tang, L. Lu, H. Li, C. Evangelinos, L. Grinberg, V. Sachdeva, G.E. Karniadakis, OpenRBC: A fast simulator of red blood cells at protein resolution. Biophys. J. **112**(10), 2030–2037 (2017)
93. A.L. Blumers, Y.-H. Tang, Z. Li, X. Li, G.E. Karniadakis, GPU-accelerated red blood cells simulations with transport dissipative particle dynamics. Comput. Phys. Commun. **217**, 171–179 (2017)
94. Z. Peng, X. Li, I.V. Pivkin, M. Dao, G.E. Karniadakis, S. Suresh, Lipid bilayer and cytoskeletal interactions in a red blood cell. Proc. Natl. Acad. Sci. U. S. A. **110**(33), 13356–13361 (2013)
95. D.A. Fedosov, B. Caswell, G.E. Karniadakis, Coarse-grained red blood cell model with accurate mechanical properties, rheology and dynamics, in *Annual International Conference of the IEEE Engineering in Medicine and Biology Society*, vol. 1–20, (2009), pp. 4266–4269
96. J.P. Hale, G. Marcelli, K.H. Parker, C.P. Winlove, P.G. Petrov, Red blood cell thermal fluctuations: Comparison between experiment and molecular dynamics simulations. Soft Matter **5**(19), 3603–3606 (2009)
97. L.-G. Jiang, H.-A. Wu, X.-Z. Zhou, X.-X. Wang, Coarse-grained molecular dynamics simulation of a red blood cell. Chin. Phys. Lett. **27**(2) (2010). Art. no. 028704
98. J. Lyu, P.G. Chen, G. Boedec, M. Leonetti, M. Jaeger, Hybrid continuum–coarse-grained modeling of erythrocytes. Comptes Rendus Mécanique **346**(6), 439–448 (2018)
99. P.B. Canham, The minimum energy of bending as a possible explanation of the biconcave shape of the human red blood cell. J. Theor. Biol. **26**(1), 61–81 (1970)
100. S. Svetina, B. Žekš, Membrane bending energy and shape determination of phospholipid vesicles and red blood cells. Eur. Biophys. J. **17**(2), 101–111 (1989)
101. H.N. Polwaththe-Gallage, S.C. Saha, E. Sauret, R. Flower, W. Senadeera, Y. Gu, SPH-DEM approach to numerically simulate the deformation of three-dimensional RBCs in non-uniform capillaries. Biomed. Eng. Online **15**, 354–370 (2016)
102. K.-i. Tsubota, S. Wada, Elastic force of red blood cell membrane during tank-treading motion: Consideration of the membrane's natural state. Int. J. Mech. Sci. **52**(2), 356–364 (2010)
103. J. Li, M. Dao, C.T. Lim, S. Suresh, Spectrin-level modeling of the cytoskeleton and optical tweezers stretching of the erythrocyte. Biophys. J. **88**(5), 3707–3719 (2005)

104. A.S. Ademiloye, L.W. Zhang, K.M. Liew, Atomistic–continuum model for probing the biomechanical properties of human erythrocyte membrane under extreme conditions. Comput. Methods Appl. Mech. Eng. **325**, 22–36 (2017)
105. G. Závodszky, B. van Rooij, V. Azizi, A. Hoekstra, Cellular level in-silico modeling of blood rheology with an improved material model for red blood cells. Front. Physiol. **8** (2017). Art no. 563
106. N.M. Geekiyanage, M.A. Balanant, E. Sauret, S. Saha, R. Flower, C.T. Lim, Y. Gu, A coarse-grained red blood cell membrane model to study stomatocyte-discocyte-echinocyte morphologies. PLoS ONE **14**(4) (2019). Art. no. 0215447
107. H.-Y. Chang, X. Li, G.E. Karniadakis, Modeling of biomechanics and biorheology of red blood cells in type 2 diabetes mellitus. Biophys. J. **113**(2), 481–490 (2017)
108. H. Delingette, Triangular springs for modeling nonlinear membranes. IEEE Trans. Vis. Comput. Graph. **14**(2), 329–341 (2008)
109. L. Miao, U. Seifert, M. Wortis, H.-G. Döbereiner, Budding transitions of fluid-bilayer vesicles: The effect of area-difference elasticity. Phys. Rev. E **49**(6), 5389–5407 (1994)
110. U. Seifert, K. Berndl, R. Lipowsky, Shape transformations of vesicles: Phase diagram for spontaneous- curvature and bilayer-coupling models. Phys. Rev. A **44**(2), 1182–1202 (1991)
111. W. Helfrich, Blocked lipid exchange in bilayers and its possible influence on the shape of vesicles. Z. Naturforsch. **29**(C), 510–515 (1974)
112. W. Helfrich, Elastic properties of lipid bilayers: Theory and possible experiments. Z. Naturforsch. **28**(11), 693–703 (1973)
113. S. Svetina, A. Ottova-Leitmannova, R. Glaser, Membrane bending energy in relation to bilayer couples concept of red blood cell shape transformations. J. Theor. Biol. **94**(1), 13–23 (1982)
114. V. Heinrich, S. Svetina, B. Žekš, Nonaxisymmetric vesicle shapes in a generalized bilayer-couple model and the transition between oblate and prolate axisymmetric shapes. Phys. Rev. E **48**(4), 3112–3123 (1993)
115. K. Khairy, J. Foo, J. Howard, Shapes of red blood cells: Comparison of 3D confocal images with the bilayer-couple model. Cell. Mol. Bioeng. **1**(2), 173–181 (2008)
116. X. Li, I.V. Pivkin, H. Liang, G.E. Karniadakis, Shape transformations of membrane vesicles from amphiphilic triblock copolymers: A dissipative particle dynamics simulation study. Macromolecules **42**(8), 3195–3200 (2009)
117. G.H.W. Lim, M. Wortis, R. Mukhopadhyay, Red blood cell shapes and shape transformations: Newtonian mechanics of a composite membrane: Sections 2.5–2.8, in *Soft Matter*, (2009), pp. 83–250
118. S. Svetina, Vesicle budding and the origin of cellular life. ChemPhysChem **10**(16), 2769–2776 (2009)
119. K. Khairy, J. Howard, Minimum-energy vesicle and cell shapes calculated using spherical harmonics parameterization. Soft Matter **7**(5), 2138–2143 (2011)
120. M. Chen, F.J. Boyle, An enhanced spring-particle model for red blood cell structural mechanics: Application to the stomatocyte–discocyte–echinocyte transformation. J. Biomech. Eng. **139**(12) (2017). Art no. 121009
121. Z.-X. Tong, X. Chen, Y.-L. He, X.-B. Liao, Coarse-grained area-difference-elasticity membrane model coupled with IB–LB method for simulation of red blood cell morphology. Physica A: Stat. Mech. Appl. **509**, 1183–1194 (2018)
122. G.H.W. Lim, A Numerical Study of Morphologies and Morphological Transformations of Human Erythrocyte Based on Membrane Mechanics, Doctor of Philosophy, Department of Physics, Simon Fraser University, (2003)
123. J.P. Mills, L. Qie, M. Dao, C.T. Lim, S. Suresh, Nonlinear elastic and viscoelastic deformation of the human red blood cell with optical tweezers. Mech. Chem. Biosyst. **1**(3), 169–180 (2004)
124. H.N. Polwaththe-Gallage, S.C. Saha, Y. Gu, Deformation of a Three-Dimensional Red Blood Cell in a Stenosed Microcapillary, Presented at the 8th Australasian Congress on Applied Mechanics (ACAM-8), Melbourne, Australia, (2014)
125. J.N. Israelachvili, *Intermolecular and Surface Forces*, 3rd edn. (Elsevier, 2011)

126. A. Iglič, V. Kralj-Iglič, H. Hägerstrand, Amphiphile induced echinocyte-spheroechinoeyte transformation of red blood cell shape. Eur. Biophys. J. **27**(4), 335–339 (1998)
127. S. Etcheverry, M.J. Gallardo, P. Solano, M. Suwalsky, O.N. Mesquita, C. Saavedra, Real-time study of shape and thermal fluctuations in the echinocyte transformation of human erythrocytes using defocusing microscopy. J. Biomed. Optics **17**(10) (2012). Art no. 106013
128. M. Nakamura, S. Bessho, S. Wada, Analysis of red blood cell deformation under fast shear flow for better estimation of hemolysis. Int. J. Numer. Methods Biomed. Eng. **30**(1), 42–54 (2014)
129. M. A. Balanant, Experimental Studies of Red Blood Cells during Storage, Doctor of Philosophy, Science and Engineering Faculty, Queensland University of Technology, (2018)
130. J. Rodriguez, T. Edeskär, S. Knutsson, Particle shape quantities and measurement techniques: A review. Electron. J. Geotech. Eng. **18**(A), 169–198 (2013)
131. S. Hénon, G. Lenormand, A. Richert, F. Gallet, A new determination of the shear modulus of the human erythrocyte membrane using optical tweezers. Biophys. J. **76**(2), 1145–1151 (1999)
132. J. Czerwinska, S.M. Wolf, H. Mohammadi, S. Jeney, Red blood cell aging during storage, studied using optical tweezers experiment. Cell. Mol. Bioeng. **8**(2), 258–266 (2015)
133. Y. Liang, Y. Xiang, J. Lamstein, A. Bezryadina, Z. Chen, Cell deformation and assessment with tunable "tug-of-war" optical tweezers, in *Conference on Lasers and Electro-Optics*, San Jose, California, 2019, Optical Society of America

Development of a Computational Modelling Platform for Patient-specific Treatment of Osteoporosis

Madge Martin, Vittorio Sansalone, and Peter Pivonka

Abstract Osteoporosis (OP) is considered as a major health burden worldwide. OP drug treatments aim at reducing the augmented risk of bone fracture caused by OP-induced loss of bone mass and increased bone matrix brittleness. The development of bone biomarkers over the past decades improved the understanding of the pathophysiology of OP, providing indicators of bone formation and resorption on a short time scale. Biomarkers can therefore be used to characterize bone remodeling and to quantify drug efficacy in OP. Recently, mechanistic pharmacokinetic-pharmacodynamic (PK-PD) models have been developed to quantitatively characterize drug effects on OP disease progression. These frameworks aim at accurately describing the mechanobiology of bone remodeling which then creates the necessary biochemical and mechanical interface for drug and exercise interventions. This chapter will present a recently-developed multiscale computational model which includes mechanobiological description of bone remodeling together with osteocyte feedback to study the effects of a virtual anti-sclerostin therapy on bone remodeling and changes in bone biomarkers. Mechanistic PK-PD models of OP treatment have great potential to quantitatively predict the long-term effects of drugs on clinical outcomes and allow patient-specific estimation of bone gain, in particular via the integration of multiple treatment options and bone mineralization.

M. Martin
School of Mechanical, Medical and Process Engineering, Queensland University of Technology, Brisbane, QLD, Australia

Laboratoire Modélisation et Simulation Multi-Echelle (MSME UMR 8208 CNRS), Université Paris-Est Créteil, Créteil, France
e-mail: madge.martin@u-pec.fr

V. Sansalone
Laboratoire Modélisation et Simulation Multi-Echelle (MSME UMR 8208 CNRS), Université Paris-Est Créteil, Créteil, France
e-mail: vittorio.sansalone@u-pec.fr

P. Pivonka (✉)
School of Mechanical, Medical and Process Engineering, Queensland University of Technology, Brisbane, QLD, Australia
e-mail: peter.pivonka@qut.edu.au

© Springer Nature Switzerland AG 2020
K. Miller et al. (eds.), *Computational Biomechanics for Medicine*,
https://doi.org/10.1007/978-3-030-42428-2_6

1 Introduction

Osteoporosis (OP) is characterised by fractures of spine, hip and wrist as primary clinical manifestations. It is a major health problem in society when considering prevalence, lifetime risk and socio-economical impact [21]. OP is also referred to as silent disease due to the fact that the disease is commonly diagnosed only after fractures occur. Understanding the origin of the disease and to find effective treatments is paramount for addressing this health issue [4]. New drugs are continually being developed and tested on animals, only the most promising ones are then tested on humans in the course of clinical trials. Drug efficacy and safety are usually assessed using biomarkers. Important biomarkers include bone mineral density (BMD) and bone turnover markers (BTMs). While the former provide insight into bone matrix properties, the latter allow to assess the activity of bone cells in the entire body by measuring bone molecular product concentrations in blood and/or urine.

Traditionally, pharmacokinetic-pharmacodynamic (PK-PD) modeling is used to characterise the time course of a drug effect with the primary objective of optimising the dosing regimen and the delivery profile. Over the past decade, PK-PD models have also been applied in the drug development process [7]. It is well known that conventional PK-PD models are descriptive, empirical and driven by large amount of data. Consequently, these models are unable to predict clinical responses beyond the data which they are based on. To overcome these limitations, more sophisticated models have been developed which take into account the underlying mechanisms of a pathology and the action of the drug, with the aim of characterising the biological processes between the drug administration and the drug effect. The latter models are referred to as mechanistic PK-PD models and rely on biomarker data and on model parameters [42, 43, 49].

As reviewed previously [54], the first model of bone remodeling coupling biochemical mechanisms was proposed by Lemaire and co-authors [25] and successively refined by Pivonka and co-workers [40, 41]. Both of these models – referred to as bone cell population models (BCPM) – take into account the major bone cell types involved in the remodeling, together with the significant regulatory factors. The major conceptual breakthrough with respect to mechanobiology of bone was achieved by Pivonka et al. to develop an evolution equation for the bone volume fraction (BV/TV) which is proportional to active osteoclast and osteoblast numbers. The latter formulation paved the way for mechanobiological extensions of the model [17, 27, 36, 39, 47]. Note that the majority of models of bone remodeling and adaptation are driven exclusively by mechanical quantities with no consideration of underlying bone-cellular activities. In this context, a major advantage of mechanistic models is to provide a translation towards bone biology and clinical bone research.

Note that both the model from Lemaire et al. and the original model from Pivonka et al. do not incorporate the concept of Frost's mechanostat [12], i.e., they lack inclusion of a mechanobiological feedback mechanism. According to the mechanostat, bone overloading leads to increased bone formation responses, whereas bone disuse leads to increased bone resorption responses. This feedback warrants that, after sufficient time, bone reaches a new equilibrium state. Given the importance of mechanobiological feedback in (re)-modeling, Pivonka and co-workers combined the bone cell population model with a micromechanical model of bone stiffness including the mechanostat concept [39, 47]. This mechanistic BCPM uses the strain energy density (SED) induced in the bone matrix as a feedback variable to control the bone formation and resorption response. The latter model was extended in several ways: (i) use of osteocyte lacunar pressure as mechanobiological feedback quantity [36] and (ii) translating SED into biochemical signalling molecules including nitric oxide and sclerostin concentration [27]. These mechanistic models of bone remodeling can further be used in combination with drug treatments to investigate drug efficacy. The following mechanistic PK-PD models have been developed: treatment of PMO with denosumab [28, 48] and treatment of an OVX rat model with PTH [55]. Moreover, Hambli et al. recently proposed a numerical finite-element study of the effects of denosumab on bone density and recovered local BMD changes in the femur observed in a human cohort [15].

The objective of this chapter is to introduce a recently developed mechanistic pharmacokinetic-pharmacodynamic (PK-PD) model of disease progression and therapeutic intervention in osteoporosis (OP) [27]. Key features of this model are: (i) bone cells concentrations (osteocytes, active osteoblasts, active osteoclasts and their precursor cells), (ii) cell-cell signalling pathways (RANK-RANKL-OPG pathway) and Wnt pathway, (iii) major regulatory molecules including parathyroid hormone (PTH), transforming growth factor β (TGF-β) and sclerostin (Scl), (iv) mechanobiological feedback, i.e., strain energy density in the bone matrix is sensed by osteoctyes to produce nitric oxide (NO) and sclerostin (Scl). This model is used to investigate postmenopausal osteoporosis (PMO) and its treatment with an anabolic drug targeting sclerostin. As a first approximation the drug action is simulated as a constant sink term in the sclerostin evolution equation. Finally, we will provide an outlook on how to extend current mechanistic models of bone remodeling in a modular fashion with the aim to continuously include new regulatory pathways and drug actions.

2 Methods

In this section, we present a mechanistic model of bone remodelling and its application towards the anabolic treatment of osteoporosis with sclerostin antibody drug treatment.

2.1 Mechanistic Tissue-Scale Model of Bone Remodeling

In the following, a description of a mechanobiological computational model of bone cell interactions in bone remodeling is provided. This model takes into account catabolic and anabolic signaling pathways including the RANK-RANKL-OPG pathway and the Wnt-Scl-LRP5/6 signaling pathway together with the action of PTH, NO, and TGF-β on bone cells as presented by Martin et al. [27]. In particular, in this mechanistic framework, one can account for the role of mechanical loading on bone cells turnover and ligands expression.

In particular, Martin et al. proposed a model focusing on two regulatory molecules produced by osteocytes which affect bone remodeling: nitric oxide and sclerostin. On the one hand, osteocytes produce high levels of nitric oxide (NO) in response to mechanical loading both in vitro and in vivo [60]. Mechanical strain stimulates NO production via the upregulation of eNOS mRNA and protein which decreases the RANKL/OPG ratio, and therefore the catabolic action of the RANK-RANKL-OPG pathway [10, 45]. On the other hand, sclerostin is a major potent mechanosensory signal inhibiting the anabolic Wnt pathway. In particular, osteocytes' sclerostin expression increases with reduced loading (mouse hindlimb unloading [46], simulated micro-gravity on osteocyte cell lines [50] or human bed rest [13]).

These mechanically-regulated processes control bone remodeling by stimulating or inhibiting ligand-binding reactions. The interactions of the aforementioned pathways are schematically illustrated in Fig. 1:

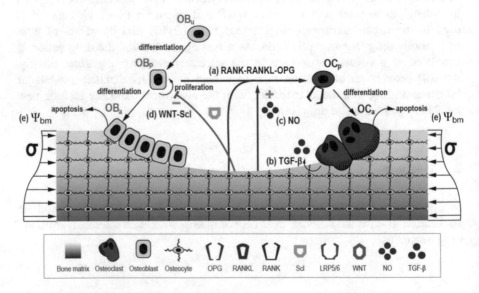

Fig. 1 Overview of the biochemical mechanostat feedback. (Adapted from Martin et al. [27])

(a) The receptor RANK is expressed on osteoclast precursor cells. RANKL, a membrane bound ligand, is expressed by osteoblast precursor cells, while OPG is produced by active osteoblasts.

(b) Action of TGF-β on bone cells is taken into account as previously described [25, 40]: TGF-β increases osteoblast precursor cells concentration (Ob$_p$) by up-regulating differentiation of uncommitted osteoblasts and down-regulating the differentiation of osteoblast precursor cells. The action of TGF-β exerted on osteoclasts is to up-regulate apoptosis of active osteoclasts

(c) Co-regulation of RANKL (expressed on osteoblast precursor cells) by NO, which is produced by osteocytes responding to changes in mechanical loading.

(d) Wnt signaling, activated through β-catenin, is an anabolic pathway promoting osteoblasts proliferation and bone formation. Extracellular Wnt ligands (produced by bone marrow stromal cells) bind to Frizzled receptor proteins and lipoprotein receptor-related proteins (LRP5/6), triggering intracellular activation of β-catenin. Sclerostin modulates the signaling pathway by interacting with LRP5/6 receptors, therefore preventing the formation of a Wnt-Frizzled-LRP5/6 complex.

(e) Osteocytes embedded in the bone matrix respond to mechanical loading which is interpreted via the mechanical stimulus Ψ_{bm}, modifying their biochemical signalling.

The bone cell types (i.e., state variables) considered in this model are: (i) osteoblast precursor cells (Ob$_p$), (ii) active osteoblasts (Ob$_a$), (iii) osteocytes (Ot), and (iv) active osteoclasts (Oc$_a$). Figure 1 displays an overview of the cell populations included in the model. The cell pools of uncommitted osteoblasts (Ob$_u$) and osteoclast precursors (Oc$_p$) are assumed to be much larger than the other cell pools and thus are not included explicitly into the model (Fig. 1).

Using the above described regulatory mechanisms, the mechanobiological model of bone remodeling can be formulated as cell balance equations describing in- and outflow of cells of the respective cell pools:

$$\frac{dOb_p}{dt} = D_{Ob_u}Ob_u\pi_{act,Ob_u}^{TGF-\beta} + P_{Ob_p}Ob_p\pi_{act,Ob_p}^{Wnt} - D_{Ob_p}Ob_p\pi_{rep,Ob_p}^{TGF-\beta} \quad (1)$$

$$\frac{dOb_a}{dt} = D_{Ob_p}Ob_p\pi_{rep,Ob_p}^{TGF-\beta} - A_{Ob_a}Ob_a \quad (2)$$

$$\frac{dOc_a}{dt} = D_{Oc_p}Oc_p\pi_{act,Oc_p}^{RANK} - A_{Oc_a}Oc_a\pi_{act,Oc_a}^{TGF-\beta} \quad (3)$$

$$\frac{dOt}{dt} = \eta\frac{df_{bm}}{dt} \quad (4)$$

where D_{Ob_u}, D_{Ob_p}, D_{Oc_p} are differentiation rates of uncommitted osteoblast progenitor cells and osteoblast/osteoclast precursor cells, respectively. P_{Ob_p} denotes the proliferation rate of osteoblast precursor cells, A_{Ob_a} is the rate of clearance of active osteoblasts through apoptosis or differentiation and A_{Oc_a} is the apoptosis rate of active osteoclasts. Equation (4) indicates that we assume that the change in osteocyte population is proportional to the change in bone matrix fraction $\frac{df_{bm}}{dt}$. The factor η indicates the average concentration of osteocytes embedded in the bone matrix In the previous set of equations (1)–(4), the populations are accounted as concentrations (numerical values in pM, as in Table 2) in the RVE. In the following, all concentrations of regulatory factors and cell numbers are evaluated with respect to the RVE.

Similar to previous model formulations, we assume that the change in bone matrix fraction (f_{bm}) depends on the number of active osteoclasts and osteoblasts (Eq. (5)) and their respective bone resorption and formation rates [38, 40, 47]:

$$\frac{df_{bm}}{dt} = -k_{res}Oc_a + k_{form}Ob_a, \tag{5}$$

where k_{res} and k_{form} are respectively the rates of bone resorption and formation (see Table 3 in Appendix).

Furthermore, differentiation, proliferation, and apoptosis are regulated by several 'activator' ($\pi_{act,X}^Y$) and 'repressor' functions ($\pi_{rep,X}^Y$), i.e., functions which promote or inhibit differentiation, proliferation or apoptosis of cells, as well as ligand production. These regulating functions are Hill functions reflecting the binding of ligands to receptors. Their mathematical definition is described in the Appendix.

2.2 Osteocyte-Driven Mechanical Feedback

As presented earlier, osteocytes biochemical signalling regulates bone remodeling. Meanwhile, this phenomenon is controlled by the mechanical environment of osteocytes, which, in turn, is affected by bone remodeling. We make the assumption here that osteocytes biochemical signalling is driven by the mechanical stimulus Ψ_{bm}, which is defined as the strain energy density in the bone matrix as per Scheiner et al. [47]. The mechanical stimulus Ψ_{bm} is calculated using a micro-mechanics framework and the Mori-Tanaka method based on Eshelby's classical matrix-inclusion problem [31]. This definition implies a dependency of the mechanical stimulus on bone's microstructure and macroscopic porosity.

This subsection develops the mathematical quantification of osteocytes biochemical response to mechanical stimuli in terms of Scl and NO production, and how these ligands then affect bone remodeling via the Wnt and RANK-RANKL-OPG pathways, respectively.

2.2.1 Modelling the Mechanical Feedback: Modulation of Osteocytes Biochemical Response

We present here the mechanostat feedback from the osteocytes, based on a setpoint strain energy criterion. Osteocytes mechanosensitivity is represented via the definition of functions regulating the production of Scl and NO by osteocytes to account for the influence of mechanical loading on osteocytes ligand production. In turn, these functions regulate the Wnt and RANKL signaling pathways. We introduce mechanically-controlled regulating functions embracing the whole range of stimuli:

- Up-regulation of osteocytes' nitric oxide production by the mechanical stimulus Ψ_{bm}: $\pi_{act,NO}^{\Psi_{bm}}$;
- Down-regulation of osteocytes' sclerostin production by the mechanical stimulus Ψ_{bm}: $\pi_{rep,Scl}^{\Psi_{bm}}$.

The feedback activator function $\pi_{act,NO}^{\Psi_{bm}}$ takes the mechanical stimulus (strain energy density in the bone matrix Ψ_{bm}) as an input and drives the osteocytes' NO production. In the same manner, $\pi_{rep,Scl}^{\Psi_{bm}}$ represents the influence of the value of the strain energy density on osteocytes' sclerostin production. Both actions are represented via sigmoidal Hill functions as suggested by Peterson et al. [37]:

$$\pi_{act,NO}^{\Psi_{bm}} = \rho_{act} + \frac{(\alpha_{act} - \rho_{act})\Psi_{bm}^{\gamma_{act}}}{\delta_{act}^{\gamma_{act}} + \Psi_{bm}^{\gamma_{act}}}, \tag{6}$$

$$\pi_{rep,Scl}^{\Psi_{bm}} = \rho_{rep} + \frac{(\alpha_{rep} - \rho_{rep})\Psi_{bm}^{\gamma_{rep}}}{\delta_{rep}^{\gamma_{rep}} + \Psi_{bm}^{\gamma_{rep}}}, \tag{7}$$

where $\rho_\sim, \alpha_\sim, \gamma_\sim, \delta_\sim$, are respectively the minimum anticipated response, the maximum anticipated response, the sigmoidicity term influencing the steepness of the response, and the value of the stimulus that produces the half-maximal response [37]. The values of the parameters defining the activator and repressor mechanical functions can be found in Table 1.

Table 1 Parameters of the mechanical regulation, as per Martin et al. [27]

Symbol	Value	Unit
ρ_{act}	0.000	–
ρ_{rep}	0.000	–
α_{act}	1.000	–
α_{rep}	1.000	–
γ_{act}	7	–
γ_{rep}	9	–
δ_{act}	$4.368 \, 10^{-6}$	–
δ_{rep}	$9.226 \, 10^{-6}$	–

2.2.2 Modelling the Action of Sclerostin on Bone Remodeling

As described in Sect. 2.1, Wnt signaling is an anabolic pathway promoting osteoblasts proliferation and bone formation. Sclerostin inhibits the Wnt pathway by binding with osteoblastic LRP5/6 receptors. In the present study, we simplify the dynamics and assume that Scl and Wnt bind directly to LRP5/6 (Fig. 1d). Utilizing this receptor-ligand binding model, Wnt signaling can be quantified by the receptor occupancy $\pi_{\text{act},\text{Ob}_p}^{\text{Wnt}}$, defined as the ratio between Wnt−LRP5/6 complexes and the total concentration of LRP5/6 receptors [LRP5/6]$_{\text{tot}}$, including the ones binding to sclerostin:

$$\pi_{\text{act},\text{Ob}_p}^{\text{Wnt}} = \frac{[\text{Wnt}-\text{LRP5/6}]}{[\text{LRP5/6}]_{\text{tot}}}, \tag{8}$$

where

$$[\text{LRP5/6}]_{\text{tot}} = [\text{LRP5/6}] + [\text{Wnt}-\text{LRP5/6}] + [\text{Scl}-\text{LRP5/6}]. \tag{9}$$

[LRP5/6] is the concentration of free LRP5/6 receptors, whereas [Wnt − LRP5/6] and [Scl−LRP5/6] are respectively the concentration of Wnt−LRP5/6 and Scl − LRP5/6 complexes expressed on osteoblast precursors.

Using the steady-state assumption, the total concentration of receptors [LRP5/6]$_{\text{tot}}$ can also be expressed as the sum of free and bound receptors as follows:

$$[\text{LRP5/6}]_{\text{tot}} = [\text{LRP5/6}] \left(1 + \frac{[\text{Wnt}]}{K_D^{\text{Wnt}-\text{LRP5/6}}} + \frac{[\text{Scl}]}{K_D^{\text{Scl}-\text{LRP5/6}}} \right), \tag{10}$$

where $K_D^{\text{Wnt}-\text{LRP5/6}}$ and $K_D^{\text{Scl}-\text{LRP5/6}}$ are the dissociation constants of the Wnt-LRP5/6 and Scl-LRP5/6 complexes, respectively.

Additionally, given that we assume that the binding reactions are much faster than the processes they regulate (steady-state assumption), there is a balance between the production and degradation of LRP5/6 (see Eq. (20) expressing the balance between ligand production and degradation in Appendix). This balance equation leads to the expression of free LRP5/6 levels as a function of LRP5/6 complexes concentrations and sclerostin levels (see Martin et al. [27] for extensive mathematical developments).

We assume that the concentration [Wnt] of available Wnt proteins stays constant, given as a basal concentration of free Wnt proteins in the medium. This assumption is based on the assumption that the degradation of the complex Wnt-LRP5/6 is negligible and on the fact that bone marrow mesenchymal stem cells (uncommitted osteoblasts Ob$_u$) are producing Wnt, while the latter population is assumed constant in our model. Additionally, the total number of LRP5/6 receptors per osteoblast precursor cell $N_{\text{Ob}_p}^{\text{LRP5/6}}$ is assumed to be constant.

The sclerostin balance (as per the general balance equation (Eq. (20))) is a function of the local sclerostin production by osteocytes which is regulated via the mechanical repressor function $\pi_{rep,Scl}^{\Psi_{bm}}$, as follows:

$$P_{Scl,b} + P_{Scl,d} = \tilde{D}_{Scl}[Scl] + \tilde{D}_{Scl-LRP5/6}[Scl - LRP5/6] \quad (11)$$

$$P_{Scl,b} = \beta_{Scl,Ot}\pi_{rep,Scl}^{\Psi_{bm}}[Ot]\left(1 - \frac{[Scl]}{[Scl]_{max}}\right) \quad (12)$$

where $P_{Scl,d}$ is an external sclerostin dosage term, which is set to zero throughout the rest of this paper, $P_{Scl,b}$ the sclerostin body production and $\tilde{D}_X$ the degradation rate of X. The set of equations (11, 12) gives the current sclerostin concentration.

Finally, the concentration of Scl obtained from the previous equations (Eqs. (11, 12)) regulates the Wnt binding to LRP5/6 (Eqs. (9, 10)), therefore driving the osteoblast precursors proliferation.

2.2.3 Modelling the Action of Nitric Oxide on Bone Remodeling

The action of nitric oxide on bone remodeling is accounted for via its regulation of the RANK-RANKL-OPG pathway. RANKL transcription is both up-regulated by PTH and inhibited by NO. These antagonistic influences were merged into a co-regulatory function $\pi_{act/rep,RANKL}^{PTH,NO}$ capturing both effects. The competition between the two actions is accounted for by the definition of the co-regulatory function as a weighted sum of the total of the activator and repressor actions and a term accounting for the combined influence:

$$\pi_{act/rep,RANKL}^{PTH,NO} = \lambda_s\left(\pi_{act,RANKL}^{PTH} + \pi_{rep,RANKL}^{NO}\right) + \lambda_c\,\pi_{act,RANKL}^{PTH}\pi_{rep,RANKL}^{NO}, \quad (13)$$

where $\lambda_s = 0.4505$ and $\lambda_c = 0.9009$ are constants respectively describing single and combined influences of respective activator/repressor functions, as per Martin et al. [27]. Note that, unlike the classical Hill functions, this competitive regulatory function $\pi_{act/rep,RANKL}^{PTH,NO}$ can take values higher than 1 ($\pi_{act/rep,RANKL}^{PTH,NO} \in [0, 1.35]$).

2.2.4 Closing the Feedback Loop of the Bone Remodeling 'mechanostat'

Using the above functions, the mechanobiological feedback loop is complete: osteocytes sense the mechanical stimulus, leading to a change in ligand production, namely Scl and NO. The latter factors act on different cells in the bone multicellular units (BMU), therefore changing the BMU remodeling response which then modifies bone matrix fraction. The change in material properties directly influences the mechanical stimulus Ψ_{bm} via the micro-mechanical representation.

2.3 Integrating Post-menopausal Osteoporosis Pathophysiology into a Mechanistic Framework of Bone Remodeling

In previous studies, increased RANKL/OPG ratios have been reported in post-menopausal osteoporotic patients [19, 26, 29], which might be resulting from the decreasing levels of estrogen which stimulates both osteoclast proliferation and activity [26]. Hence, we simulate here post-menopausal osteoporosis by means of an external injection of RANKL via an external production term in the RANKL balance (Eq. (20)): $P_{\text{RANKL},d} = 2.000\,\text{pM.day}^{-1}$.

Furthermore, studies have shown an increase of serum sclerostin in post-menopausal subjects [3, 20], while sclerostin expression (local mRNA levels) was found to decrease in animal models of menopause [20]. This discrepancy between the serum levels and the local expression of sclerostin is acknowledged by assuming an exponential decay of the degradation rate of sclerostin: $\tilde{D}_{\text{Scl}}(t = t_{menop} + \tau) = \tilde{D}_{\text{Scl,PMO}}(\tau)$, where $\tilde{D}_{\text{Scl,PMO}}$ is the function defined as follows:

$$\tilde{D}_{\text{Scl,PMO}}(\tau) = \tilde{D}_{\text{Scl}}^0 \exp(-\frac{\tau}{\tau_{PMO}}), \tag{14}$$

where $\tau_{PMO} = 20\,\text{yrs}$ is the characteristic time of the decay.

2.4 Modelling Osteoporosis Treatment with Anti-sclerostin Antibody Therapy: From Drug Pharmacokinetics to the Quantification of Bone Remodeling

As described earlier, osteocytes produce sclerostin, and the latter binds to low-density lipoprotein receptor-related protein (LRP), therefore inhibiting Wnt signaling and the anabolic β-catenin pathway [5]. Therefore, sclerostin levels are directly connected to bone turnover and are negatively correlated to bone formation.

In the past decade, several studies assessed the influence of sclerostin monoclonal antibody, inhibiting sclerostin regulation of bone formation. Warmington and collaborators led in 2004 the first research testing the therapeutic potential of sclerostin neutralizing antibodies [56]. They found that a sclerostin monoclonal antibody mediated blockade led to a significant BMD increase in adult mice and rats, including up to 64% in tibial metaphysis trabecular bone. Since then, various studies investigated the administration of a sclerostin monoclonal antibody as a means to counter osteoporosis-induced bone loss [22, 35] or promote bone fracture healing [22, 23].

A novel anabolic treatment for osteoporosis has been developed in the recent years which is based on an anti-sclerostin antibody commercialized under the name Evenity® [2]. This drug treatment is approved for marketing in Japan, South Korea,

US and Canada, and recently in EU. Evenity® treatment relies on a humanized anti-sclerostin antibody, called romosozumab. While the antibody injections have a short-term significantly positive impact on bone mass [30, 34], the influence of the treatment on bone mass and turnover markers [51, 59] as well as its interaction with other pathways [53] are not fully understood.

Pharmacokinetic (PK) models help understand how drugs interact with the biological systems, in particular in the case of monoclonal antibodies [6, 8]. These models can be used to study drug pharmacodynamics (PD), which gives insights on the biochemical, physiological and molecular effects of drugs on the body. Studying sclerostin antibody pharmacokinetics and pharmacodynamics has the potential to shed light on the dynamics of bone resorption and formation during a treatment (see Eudy et al.'s study on romosozumab [9] and Tang et al. on blosozumab [52]).

Pharmacokinetic modelling involves the choice of model compartments. Generally, the central compartment refers to the compartment where the drug binds to the target (e.g., antibody binding to the ligand in the serum). The addition of a depot accounts for a time-delay between drug administration and the absorption of the drug into the central compartment, for example to account for the injection mode: intravenous injections introduce the drug directly into the central compartment, whereas drugs injected subcutaneously are continuously absorbed into the serum [9, 48]. The addition of supplementary compartments can account for large time delays (lymph compartment), or for processes that occur in select sites after the binding process in the central compartment. For instance, in the context of anti-sclerostin drug treatment, Eudy et al. chose to integrate a peripheral compartment in their PK model of romosozumab [9], implying a slower distribution in bone tissue than in the central compartment.

The definition of the PK model allows one to follow drug levels in the model compartment(s). The drug concentration is obtained as a function of time depending on dosage and treatment interval. In two-compartment models, the tissue and central compartments are separate and the evolution of drug concentration depends on the current target concentration, while, in one-compartment models, the current target concentration does not affect drug kinetics. In the context of romosozumab pharmacokinetics, the drug enters the sclerostin balance as another binding molecule. Therefore, Eq. (11) becomes:

$$P_{Scl,b} + P_{Scl,d} =$$
$$\tilde{D}_{Scl}[Scl] + \tilde{D}_{Scl-LRP5/6}[Scl - LRP5/6] + \tilde{D}_{Scl-Rom}[Scl - Rom], \quad (15)$$

where Rom is the concentration of romosozumab.

As a result, pharmacokinetics can be linked to bone remodelling via the biochemical regulation of bone multicellular units.

3   Results

In this section, we illustrate the ability of a mechanistic representation of bone remodeling to estimate bone cells dynamics and bone turnover in the context of osteoporosis, and in particular in the case of an anti-sclerostin anabolic treatment with romosozumab.

3.1   Osteocytes Anabolic and Catabolic Responses to Mechanical Stimuli

As explained in Sect. 2.2, where the Scl and NO regulatory functions have been introduced, mechanobiological regulations via biochemical feedback are C^1 continuous functions depending on Ψ_{bm}, which may differ from the homeostatic stimulus $\breve{\Psi}_{bm}$, depending on bone tissue properties and loading conditions.

3.1.1   Influence of Mechanically-Controlled Signalling Pathways in PMO

We investigated the contribution of different signaling pathways involved in mechanobiological feedback on changes of the trabecular bone matrix fraction (f_{bm}) (Fig. 2). For this purpose, we activated/inactivated the NO production term and/or the Wnt activator function. For instance, in order to disable the NO pathway, we artificially kept the osteocyte NO production rate constant (no regulation: $\forall t, \pi_{act,NO}^{\Psi_{bm}}(\Psi_{bm}(t)) = \pi_{act,NO}^{\Psi_{bm}}(\breve{\Psi}_{bm}))$, corresponding to the initial homeostatic state. This allowed to visualize what part the catabolic NO regulation played in the remodeling. In the same way, we were able to disable the Scl pathway by maintaining the regulatory function $\pi_{rep,Scl}^{\Psi_{bm}}$ constant, in order to analyze the role of Wnt signaling in the mechanobiological feedback. This strategy was applied to a simulation of post-menopausal osteoporosis and the quantification of bone loss in the forearm (Fig. 2a). Initial values of the simulation correspond to steady-state values, which are displayed in Table 2.

In Fig. 2a, the thin solid line represents the absence of mechanical feedback, i.e., no model response to changes in mechanical loading. The thick solid line represents both the anabolic and catabolic pathways active in the bone remodeling model, which corresponds to the complete model. The dash-dotted and dashed lines respectively represent the cases where the Wnt/Scl-pathway feedback and the NO-pathway feedback are disabled, respectively.

One may note that the impact of NO production on the overall bone response due to osteoporosis is small, which indicates that the catabolic feedback is secondary in the mechanostat feedback in an osteoporotic state. On the other hand, when setting

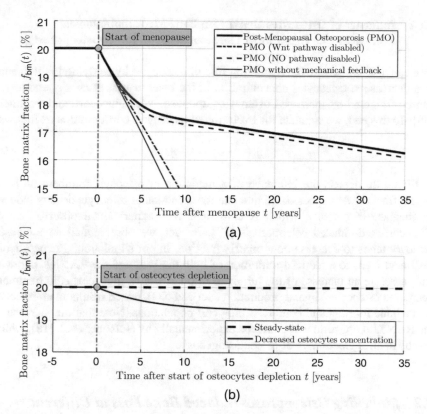

Fig. 2 Influence of active/inactive signaling pathways regulating mechanobiological feedback in remodeling response: simulation of (**a**) post-menopausal osteoporosis (PMO) and (**b**) decrease in osteocytes numbers

Table 2 Steady-state values of bone cell concentrations and tissue-scale stress σ for bone remodeling simulations at the forearm, as per Martin et al. [27]

Symbol	Value	Unit
Ob_a^0	$8.939 \ 10^{-4}$	pM
Ob_p^0	$1.129 \ 10^{-3}$	pM
Ob_u^0	$1.000 \ 10^{-2}$	pM
Oc_a^0	$1.788 \ 10^{-5}$	pM
Oc_p^0	$5.592 \ 10^{-3}$	pM
σ_{FA}	-3.350	MPa

the sclerostin expression to a constant value, bone loss increases significantly, meaning that the mechanical feedback is weaker. This suggests that osteocyte sclerostin production drives the anabolic feedback response, while NO only has a minor contribution.

3.1.2 Influence of Osteocytes Density on Bone Mechanobiological Feedback: The Example of Glucocorticoid-Induced Osteoporosis

One can also investigate the regulating role of osteocytes by studying the influence of a decrease in osteocytes concentration in the bone matrix. Such a phenomenon can be triggered by increased osteocytes apoptosis with glucocorticoid treatment [58]. To this end, we simulate the evolution of osteocytes concentration as follows:

$$\eta_t(t \geq 0) = 0.9\eta. \tag{16}$$

Figure 2b depicts the influence of a depletion of osteocyte numbers on bone matrix fraction. As expected, we find that the reduction of osteocyte density induces an increase in porosity, which is consistent with experimental observations of glucocorticoid-induced osteoporosis [57]. In fact, we observe that an increased turnover tends to decrease bone matrix fraction. In our simulation, glucocorticoid treatment leads to a reduced inhibition of both the Wnt and the RANKL pathway and therefore an increase of the turnover. In other words, glucocorticoid treatment decreases osteocytes' ligand production (Scl and NO), which results in an increase in the bone forming and bone resorbing cell populations. Note that the increase of the RANKL/OPG ratio was observed experimentally by Hofbauer et al. [18], which results in an augmentation of osteoclastogenesis.

3.2 Modelling Osteoporosis-Induced Bone Loss at Different Bone Sites

We compared our model of post-menopausal osteoporosis to several existing longitudinal studies in human at different bone sites [1, 11, 14, 16, 33, 44]. Note that each bone site is characterized by a different value of the bone matrix fraction f_{bm}, and therefore a specific value of the habitual stress σ_{ss}.

While acknowledging that the experimental measurements of BMD do not reflect the exact evolution of the bone matrix fraction, we assume here that they are close enough for us to compare their trend to our simulations.

The values of the stresses in Table 2 relate to the mechanical environment in the forearm ($f_{bm} = 20\%$). Now, based on data from the literature, we assumed respectively for the femoral neck and the lumbar spine that their steady-state bone matrix fractions were $f_{bm} = 25\%$ [32] and $f_{bm} = 12.5\%$ [24], which corresponds to habitual stresses $\sigma_{FN} = -4.405\,\text{MPa}$ and $\sigma_{LS} = -2.041\,\text{MPa}$.

Figure 3 displays literature experimental results for the evolution of BMD with time in post-menopausal osteoporosis at different bone sites ((a) distal radius, (b) lumbar spine and (c) femoral neck), along with our simulation results (solid lines). While the experimental data exhibits large standard deviations, the model is able to predict the mean trends providing confidence in the presented model formulation.

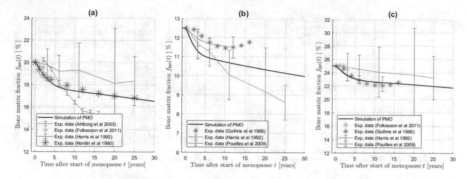

Fig. 3 Evolution of the bone matrix fraction (f_{bm}) with time: comparison of the simulations in (**a**) distal radius, (**b**) lumbar spine and (**c**) femoral neck with experimental results on the evolution of BMD in post-menopausal osteoporosis. (Reproduced from Martin et al. [27], with permission)

3.3 Virtual Anti-sclerostin Therapy of Post-menopausal Osteoporosis

In this Subsection, we study the effect of an anti-sclerostin antibody injection (romosozumab). Here, we do not account for pharmacokinetics and investigate the impact of the drug on bone turnover markers and bone matrix fraction (f_{bm}). To this end, we simulate post-menopausal osteoporosis (PMO) as per Sect. 2.3 and we simulate a sclerostin antibody injection after 5 years of PMO by setting romosozumab concentration as follows:

$$[\text{Rom}](t) = \begin{cases} 2 \cdot 10^{-1}\,\text{pM} & \text{if } t > 0 \,\&\, t < 1\,\text{month}, \\ 0 & \text{otherwise}. \end{cases} \tag{17}$$

Figure 4 describes the changes in bone remodeling biomarkers following the drug injection. In particular, one can observe that cell numbers increase in different proportions after injection (52% increase in osteoblasts numbers against 8.2% increase in osteoclasts numbers after 1 month) and decreases after the treatment stops (a). One may also notice that the model captures the delay between the primary osteoclasts response and the osteoblasts proliferation that happens afterwards.

As expected, the percentage of LRP5/6 receptor occupancy by Wnt proteins (evaluated through the Hill function $\pi_{\text{act,Ob}_p}^{\text{Wnt}}$) increases (b), as a consequence of low sclerostin serum levels (d). One may also note the feedback of the system which increases osteocytes sclerostin expression (c) in consequence of the higher bone matrix fraction resulting from drug injection (b). Conversely, before injection, sclerostin expression decreases: osteocytes respond anabolically to the PMO-induced bone loss. During this simulation, the variations of nitric oxide in terms of expression by osteocytes and serum levels stay small in comparison to that of

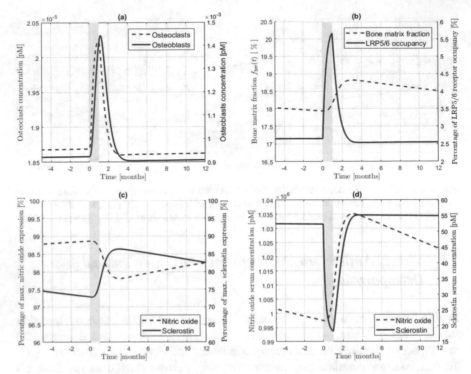

Fig. 4 Evolution of osteoclasts (*red, dashed*) and osteoclasts (*blue, solid*) concentration (**a**), bone matrix volume fraction f_{bm} (*red, dashed*) and LRP5/6 occupancy by Wnt proteins $\pi^{\text{Wnt}}_{\text{act,Ob}_p}$ (*blue, solid*) (**b**), percentage of maximum nitric oxide (*red, dashed*) and sclerostin (*blue, solid*) expression by osteocytes (**c**) and their serum levels (**d**) after an imposed change of romosozumab levels ([Rom] $= 2 \cdot 10^{-1}$ pM, $t < 1$ month (*gray area*))

sclerostin (c,d). This last observation is consistent with the findings highlighted in Sect. 3.1, as PMO is mainly controlled by the anabolic sclerostin pathway.

To summarize, the present model gives the tools for a comprehensive understanding of bone remodeling during romosozumab treatment by providing a description of biochemical changes in bone tissue.

4 Discussion and Outlook

The presented numerical simulation results for postmenopausal osteoporosis and its intervention with a sclerostin inhibitor have demonstrated that the proposed model is able to describe disease progression accurately. The virtual drug intervention strategy that was applied, i.e., targeting osteocyte produced sclerostin, seems very effective in stopping bone loss and eventually restoring bone mass. The simulations also showed that inclusion of a mechanical feedback system is essential for simulations of longer time scales in order to achieve good comparison with experimental data. The presented BCPM is an extension of earlier models of bone cell populations that describe the bone remodeling process. A major feature of the model is incorporation of osteocytes which are crucial for the formulation of mechanobiological feedback. The latter was achieved via the production of sclerostin and nitric oxide by osteocytes that, respectively, inhibit osteoblastogenesis and the catabolic RANK-RANKL-OPG pathway. As was pointed out in Martin et al., the model simulations reflect essential features of the bone remodeling process [27]:

- The BCPM describes the dynamics of bone remodeling accounting for changes in mechanical loading, as well as hormonal changes (post-menopausal osteo-porosis).
- Anabolic and catabolic bone remodeling responses were separated in the model based on two different signaling pathways.
- Catabolic model responses were linked to the nitric oxide (NO) pathway. This model feature was capable of driving resorptive osteoclastic activity. Hence, NO production by osteocytes was connected to bone loss as well as achievement of a new steady state.
- Conversely, anabolic model responses were linked to the Wnt signaling pathway. The latter model feature is a key factor for the observed bone formation response. Furthermore, it helps stabilizing the bone matrix fraction in PMO as it counters the enhanced resorptive action of the RANKL pathway.

Patient-specific modelling requires the quantification of bone biomarkers which helps assessing the state of the disease, by quantifying resorption and apposition (turnover markers) or quantifying site-specific bone loss (BMD). This data provides the tools to generate a targeted bone gain: bone turnover markers indicate the impor-tance of signalling pathways and assessment of BMD at different time points marks the evolution of bone tissue. By tuning the influence of the signalling pathways – via the corresponding Hill coefficients – and the severity of the PMO-induced bone loss – via PMO parameter $P_{RANKL,d}$, one could account for variations among patients. This method could allow for the design of a specific drug treatment plan to obtain targeted bone gains, provided a prior validation of the pharmacodynamics of the drug, in particular via its impact on BMD and bone turnover markers. Note that the presented numerical modelling platform involves solving a system of six differential equations (Eqs. (1)–(4)) and one algebraic equation corresponding to RANKL balance. This operations only requires short computation time (less than 30 s for a simulation of PMO (Fig. 3)). This would eventually enable easy application to clinics in the form of patient-specific numerical calculations informing treatment planning.

As a future outlook, we envision that a computational modeling platform for osteoporosis and various intervention strategies will be developed where we can share our models with the wider bone research community. Particularly, the modular structure of BCPM may allow to continuously extend various model features with latest discoveries of signalling pathways and regulatory factors in bone cells involved in the bone remodeling process. Also, the platform will allow to develop a suit of pharmacokinetic (PK) models that target various signalling molecules in the BCPM. The interventions that can be studied are physiological exercise and drug treatments including denosumab, PTH and others. Ultimately, use of this platform will help understanding complex interactions in multidrug treatments of OP and come up with new hypothesis for effective combinations of currently available OP drugs.

Appendix

In line with Pivonka et al. [40], the balance of the production rate P_L of a ligand L with its degradation D_L – which itself can be assumed to be proportional to the concentration of L – reads as follows:

$$P_L + D_L = P_L - (\tilde{D}_L[L] + \sum_S \tilde{D}_{L-S}[L - S]) = 0, \tag{18}$$

where $\tilde{D}_Y$ is the degradation rate of the species Y and $[L - S]$ represents the concentration of ligand L bound to S, a species in the medium that can bind to L. Note that the degradation D_L comprises the degradation of the ligand in all its forms, including bound complexes L-S. The parameters regulating the ligand balance are listed in Table 3.

The production rate can be decomposed into two components: $P_{L,b}$ (body production) and $P_{L,d}$ (external dosage). The body production is assumed to be limited by a maximum concentration $[L]_{\max}$, leading to the following expression:

$$P_{L,b} = \sum_{X,Y} \beta_{L,X} \pi^Y_{\text{act/rep},X} X (1 - \frac{[L]}{[L]_{\max}}), \tag{19}$$

$$P_{L,b} + P_{L,d} = \tilde{D}_L[L] + \sum_S \tilde{D}_{L-S}[L - S], \tag{20}$$

where X is the concentration of the cell type X producing L, with a production rate $\beta_{L,X}$ regulated by the species Y by means of the regulating activator or repressor function $\pi^Y_{\text{act/rep},X}$.

Moreover, TGF-β levels were calculated as derived in [40]: $[\text{TGF} - \beta] = \alpha \, \text{Oc}_a$, where α is a parameter listed in Table 3.

Pivonka et al. [38] formalized the regulation of a cell population X, and in particular through their proliferation, differentiation or ligand production. The regulation via the formation of the complex $L - R$ is quantitatively defined as the ratio between the occupied receptors R by ligands L and the total number of receptors:

Symbol	Value	Unit
Table 3 Biochemical parameters regulating bone remodeling, as per Martin et al. [27]		
Differentiation rates		
D_{Ob_u}	$1.660 \ 10^{-1}$	day^{-1}
D_{Ob_p}	$1.850 \ 10^{-1}$	day^{-1}
D_{Oc_p}	$1.958 \ 10^{-2}$	day^{-1}
Proliferation rate		
P_{Ob_p}	2.203	day^{-1}
Clearance rates		
A_{Ob_a}	$2.120 \ 10^{-1}$	day^{-1}
A_{Oc_a}	10.00	day^{-1}
Release factor of TGF-β		
α	1.000	–
Resorption/Formation rates		
k_{res}	2500	$pM^{-1}.day^{-1}$
k_{form}	50.00	$pM^{-1}.day^{-1}$
Dissociation coefficients of Hill functions		
$K_{D,act}^{TGF-\beta}$	$5.633 \ 10^{-4}$	pM
$K_{D,rep}^{TGF-\beta}$	$1.754 \ 10^{-4}$	pM
$K_{D,act}^{RANK}$	16.70	pM
Concentrations of osteocytes in bone matrix		
η	$4.143 \ 10^{-8} \ (4.143 \ 10^{-2})$	$pmol.mm^{-1}$ (pM)

$$\pi_{act,X}^{L} = \frac{[L-R]}{[R]_{tot}} = \frac{[L-R]}{[R] + \sum_{L'}[L'-R]}, \tag{21}$$

where $[L-R]$ is the concentration of ligands bound to the receptor R, and L' is any ligand that can bind to the receptor R (including L).

Conversely, the repressor action of the receptor-ligand binding reads:

$$\pi_{rep,X}^{L} = \frac{[R]_{tot} - [L-R]}{[R]_{tot}} = \frac{[R] + \sum_{L' \neq L}[L'-R]}{[R] + \sum_{L'}[L'-R]} \tag{22}$$

In the context of a simple ligand-receptor binding without competition, the above expressions result in simple first-order Hill activator and repressor functions.

References

1. H.G. Ahlborg, O. Johnell, C.H. Turner, G. Rannevik, M.K. Karlsson, Bone loss and bone size after menopause. N. Engl. J. Med. **349**(4), 327–334 (2003). https://doi.org/10.1056/NEJMoa022464

2. Amgen Inc.: Evenity [Prescribing Information]. Tech. rep. (Amgen Inc., Thousand Oaks, California, 2019). https://www.pi.amgen.com/~/media/amgen/repositorysites/pi-amgen-com/ evenity/evenity_pi_hcp_english.ashx
3. M.S.M. Ardawi, H.A. Al-Kadi, A.A. Rouzi, M.H. Qari, Determinants of serum sclerostin in healthy pre- and postmenopausal women. J. Bone Miner. Res. 26(12), 2812–2822 (2011). https://doi.org/10.1002/jbmr.479
4. D.M. Black, C.J. Rosen, Postmenopausal osteoporosis. N. Engl. J. Med. 374(3), 254–262 (2016). https://doi.org/10.1056/NEJMcp1513724. http://www.nejm.org/doi/10.1056/ NEJMcp1513724
5. L.F. Bonewald, M.L. Johnson, Osteocytes, mechanosensing and Wnt signaling. Bone 42(4), 606–615 (2008). https://doi.org/10.1016/j.bone.2007.12.224
6. Y. Cao, W.J. Jusko, Incorporating target-mediated drug disposition in a minimal physiologically-based pharmacokinetic model for monoclonal antibodies. J. Pharmacokinet. Pharmacodyn. 41(4), 375–387 (2014). https://doi.org/10.1007/s10928-014-9372-2
7. M. Danhof, J. de Jongh, E.C.M. De Lange, O. Della Pasqua, B.A. Ploeger, R.A. Voskuyl, Mechanism-based pharmacokinetic-pharmacodynamic modeling: biophase distribution, receptor theory, and dynamical systems analysis. Annu. Rev. Pharmacol. Toxicol. 47, 357–400 (2007). https://doi.org/10.1146/annurev.pharmtox.47.120505.105154
8. P. Dua, E. Hawkins, P. van der Graaf, A tutorial on target-mediated drug disposition TMDD models. CPT Pharmacometrics Syst. Pharmacol. 4(6), 324–337 (2015). https://doi.org/10. 1002/psp4.41
9. R. Eudy, M. Gastonguay, K. Baron, M. Riggs, Connecting the dots: linking osteocyte activity and therapeutic modulation of sclerostin by extending a multiscale systems model. CPT Pharmacometrics Syst. Pharmacol. 4(9), 527–536 (2015). https://doi.org/10.1002/psp4.12013
10. X. Fan, E. Roy, L. Zhu, T.C. Murphy, C. Ackert-Bicknell, C.M. Hart, C. Rosen, M.S. Nanes, J. Rubin, Nitric oxide regulates receptor activator of nuclear factor-κB ligand and osteoprotegerin expression in bone marrow stromal cells. Endocrinology 145(2), 751–759 (2004). https://doi. org/10.1210/en.2003-0726
11. J. Folkesson, J. Goldenstein, J. Carballido-Gamio, G. Kazakia, A.J. Burghardt, A. Rodriguez, R. Krug, A.E. de Papp, T.M. Link, S. Majumdar, Longitudinal evaluation of the effects of alendronate on MRI bone microarchitecture in postmenopausal osteopenic women. Bone 48(3), 611–621 (2011). https://doi.org/10.1016/j.bone.2010.10.179
12. H.M. Frost, Bone "mass" and the "mechanostat": a proposal. Anat. Rec. 219(1), 1–9 (1987). https://doi.org/10.1002/ar.1092190104
13. A. Gaudio, P. Pennisi, C. Bratengeier, V. Torrisi, B. Lindner, R.A. Mangiafico, I. Pulvirenti, G. Hawa, G. Tringali, C.E. Fiore, Increased sclerostin serum levels associated with bone formation and resorption markers in patients with immobilization-induced bone loss. J. Clin. Endocrinol. Metab. 95(5), 2248–2253 (2010). https://doi.org/10.1210/jc.2010-0067
14. J.R. Guthrie, P.R. Ebeling, J.L. Hopper, E. Barrett-Connor, L. Dennerstein, E.C. Dudley, H.G. Burger, J.D. Wark, A prospective study of bone loss in menopausal Australian-born women. Osteoporos. Int. 8(3), 282–290 (1998). https://doi.org/10.1007/s001980050066
15. R. Hambli, M.H. Boughattas, J.L. Daniel, A. Kourta, Prediction of denosumab effects on bone remodeling: a combined pharmacokinetics and finite element modeling. J. Mech. Behav. Biomed. Mater. 60, 492–504 (2016). https://doi.org/10.1016/j.jmbbm.2016.03.010. http:// www.sciencedirect.com/science/article/pii/S1751616116300303
16. S. Harris, B. Dawson-Hughes, Rates of change in bone mineral density of the spine, heel, femoral neck and radius in healthy postmenopausal women. Bone Miner. 17(1), 87–95 (1992). https://doi.org/10.1016/0169-6009(92)90713-N
17. P.L. Hitchens, P. Pivonka, F. Malekipour, R.C. Whitton, Mathematical modelling of bone adaptation of the metacarpal subchondral bone in racehorses. Biomech. Model. Mechanobiol. 17(3), 877–890 (2018). https://doi.org/10.1007/s10237-017-0998-z. http://link.springer.com/ 10.1007/s10237-017-0998-z
18. L.C. Hofbauer, C.A. Kühne, V. Viereck, The OPG/RANKL/RANK system in metabolic bone diseases. J. Musculoskelet. Neuronal Interact. 4(3), 268–275 (2004)

19. S. Jabbar, J. Drury, J.N. Fordham, H.K. Datta, R.M. Francis, S.P. Tuck, Osteoprotegerin, RANKL and bone turnover in postmenopausal osteoporosis. J. Clin. Pathol. **64**(4), 354–357 (2011). https://doi.org/10.1136/jcp.2010.086595

20. S. Jastrzebski, J. Kalinowski, M. Stolina, F. Mirza, E. Torreggiani, I. Kalajzic, H.Y. Won, S.K. Lee, J. Lorenzo, Changes in bone sclerostin levels in mice after ovariectomy vary independently of changes in serum sclerostin levels. J. Bone Miner. Res. **28**(3), 618–626 (2013). https://doi.org/10.1002/jbmr.1773

21. J. Kanis, C. Cooper, R. Rizzoli, J.Y. Reginster, European guidance for the diagnosis and management of osteoporosis in postmenopausal women. Osteoporos. Int. **30**(1), 3–44 (2019). https://doi.org/10.1007/s00198-018-4704-5. http://link.springer.com/10.1007/s00198-018-4704-5

22. D. Ke, D. Padhi, C. Paszty, Sclerostin is an important target for stimulating bone formation, restoring bone mass and enhancing fracture healing. Bone **46**, S15 (2010). https://doi.org/10.1016/j.bone.2010.01.022

23. B. Kruck, E.A. Zimmermann, S. Damerow, C. Figge, C. Julien, D. Wulsten, T. Thiele, M. Martin, R. Hamdy, M.K. Reumann, G.N. Duda, S. Checa, B.M. Willie, Sclerostin neutralizing antibody treatment enhances bone formation but does not rescue mechanically induced delayed healing. J. Bone Miner. Res. **33**(9), 1686–1697 (2018). https://doi.org/10.1002/jbmr.3454

24. E. Legrand, D. Chappard, C. Pascaretti, M. Duquenne, S. Krebs, V. Rohmer, M.F. Basle, M. Audran, Trabecular bone microarchitecture, bone mineral density, and vertebral fractures in male osteoporosis. J. Bone Miner. Res. **15**(1), 13–19 (2000). https://doi.org/10.1359/jbmr.2000.15.1.13

25. V. Lemaire, F.L. Tobin, L.D. Greller, C.R. Cho, L.J. Suva, Modeling the interactions between osteoblast and osteoclast activities in bone remodeling. J. Theor. Biol. **229**(3), 293–309 (2004). https://doi.org/10.1016/j.jtbi.2004.03.023

26. U. Lerner, Bone remodeling in post-menopausal osteoporosis. J. Dent. Res. **85**(7), 584–595 (2006). https://doi.org/10.1177/154405910608500703. http://journals.sagepub.com/doi/10.1177/154405910608500703

27. M. Martin, V. Sansalone, D.M.L. Cooper, M.R. Forwood, P. Pivonka, Mechanobiological osteocyte feedback drives mechanostat regulation of bone in a multiscale computational model. Biomech. Model. Mechanobiol. **18**(5), 1475–1496 (2019). https://doi.org/10.1007/s10237-019-01158-w. http://link.springer.com/10.1007/s10237-019-01158-w

28. J. Martínez-Reina, P. Pivonka, Effects of long-term treatment of denosumab on bone mineral density: insights from an in-silico model of bone mineralization. Bone **125**, 87–95 (2019). https://doi.org/10.1016/j.bone.2019.04.022. https://linkinghub.elsevier.com/retrieve/pii/S8756328219301607

29. M. McClung, Role of RANKL inhibition in osteoporosis. Arthritis Res. Ther. **9**(Suppl. 1), 1–6 (2007). https://doi.org/10.1186/ar2167

30. M.R. McClung, J.P. Brown, A. Diez-Perez, H. Resch, J. Caminis, P. Meisner, M.A. Bolognese, S. Goemaere, H.G. Bone, J.R. Zanchetta, J. Maddox, S. Bray, A. Grauer, Effects of 24 months of treatment with romosozumab followed by 12 months of denosumab or placebo in postmenopausal women with low bone mineral density: a randomized, double-blind, phase 2, parallel group study. J. Bone Miner. Res. **33**(8), 1397–1406 (2018). https://doi.org/10.1002/jbmr.3452

31. T. Mori, K. Tanaka, Average stress in matrix and average elastic energy of materials with misfitting inclusions. Acta Metall. **21**(5), 571–574 (1973). https://doi.org/https://doi.org/10.1016/0001-6160(73)90064-3. http://www.sciencedirect.com/science/article/pii/0001616073900643

32. A. Nazarian, J. Muller, D. Zurakowski, R. Müller, B.D. Snyder, Densitometric, morphometric and mechanical distributions in the human proximal femur. J. Biomech. **40**(11), 2573–2579 (2007). https://doi.org/10.1016/j.jbiomech.2006.11.022

33. B.E.C. Nordin, A.G. Need, B.E. Chatterton, M. Horowitz, H.A. Morris, The Relative contributions of age and years since menopause to postmenopausal bone loss. J. Clin. Endocrinol. Metab. **70**(1), 83–88 (1990). https://doi.org/10.1210/jcem-70-1-83

34. D. Padhi, M. Allison, A.J. Kivitz, M.J. Gutierrez, B. Stouch, C. Wang, G. Jang, Multiple doses of sclerostin antibody romosozumab in healthy men and postmenopausal women with low bone mass: a randomized, double-blind, placebo-controlled study. J. Clin. Pharmacol. **54**(2), 168–178 (2014). https://doi.org/10.1002/jcph.239

35. D. Padhi, B. Stouch, G. Jang, L. Fang, M. Darling, H. Glise, M.K. Robinson, S.S. Harris, E. Posvar, Anti-Sclerostin antibody increases markers of bone formation in healthy post-menopausal women. J. Bone Miner. Res. **22**(1), S37–S37 (2007)

36. M.I. Pastrama, S. Scheiner, P. Pivonka, C. Hellmich, A mathematical multiscale model of bone remodeling, accounting for pore space-specific mechanosensation. Bone **107**, 208–221 (2018). https://doi.org/10.1016/j.bone.2017.11.009. http://linkinghub.elsevier.com/retrieve/pii/S8756328217304271

37. M.C. Peterson, M.M. Riggs, A physiologically based mathematical model of integrated calcium homeostasis and bone remodeling. Bone **46**(1), 49–63 (2010). https://doi.org/10.1016/j.bone.2009.08.053

38. P. Pivonka, P.R. Buenzli, C.R. Dunstan, *A Systems Approach to Understanding Bone Cell Interactions in Health and Disease* (Cell Interaction, Sivakumar Gowder, IntechOpen, 2012). https://doi.org/10.5772/51149. Available from: https://www.intechopen.com/books/cell-interaction/a-systems-approach-to-understanding-bone-cell-interactions-in-health-and-disease

39. P. Pivonka, P.R. Buenzli, S. Scheiner, C. Hellmich, C.R. Dunstan, The influence of bone surface availability in bone remodelling – a mathematical model including coupled geometrical and biomechanical regulations of bone cells. Eng. Struct. **47**, 134–147 (2013)

40. P. Pivonka, J. Zimak, D.W. Smith, B.S. Gardiner, C.R. Dunstan, N.A. Sims, T.J. Martin, G.R. Mundy, Model structure and control of bone remodeling: a theoretical study. Bone **43**(2), 249–263 (2008). https://doi.org/10.1016/j.bone.2008.03.025

41. P. Pivonka, J. Zimak, D.W. Smith, B.S. Gardiner, C.R. Dunstan, N.a. Sims, T.J. Martin, G.R. Mundy, Theoretical investigation of the role of the RANK-RANKL-OPG system in bone remodeling. J. Theor. Biol. **262**(2), 306–316 (2010). https://doi.org/10.1016/j.jtbi.2009.09.021

42. T.M. Post, J.I. Freijer, J. DeJongh, M. Danhof, Disease system analysis: Basic disease progression models in degenerative disease. Pharm. Res. **22**(7), 1038–1049 (2005). https://doi.org/10.1007/s11095-005-5641-5

43. T.M. Post, S. Schmidt, L.A. Peletier, R. de Greef, T. Kerbusch, M. Danhof, Application of a mechanism-based disease systems model for osteoporosis to clinical data. J. Pharma-cokinet. Pharmacodyn. **40**(2), 143–156 (2013). https://doi.org/10.1007/s10928-012-9294-9. http://www.ncbi.nlm.nih.gov/pubmed/23315144

44. J. Pouilles, F. Tremollieres, C. Ribot, Effect of menopause on femoral and vertebral bone loss. J. Bone Miner. Res. **10**(10), 1531–1536 (2009). https://doi.org/10.1002/jbmr.5650101014

45. J. Rahnert, X. Fan, N. Case, T.C. Murphy, F. Grassi, B. Sen, J. Rubin, The role of nitric oxide in the mechanical repression of RANKL in bone stromal cells. Bone **43**(1), 48–54 (2008). https://doi.org/10.1016/j.bone.2008.03.006

46. A.G. Robling, P.J. Niziolek, L.A. Baldridge, K.W. Condon, M.R. Allen, I. Alam, S.M. Mantila, J. Gluhak-Heinrich, T.M. Bellido, S.E. Harris, C.H. Turner, Mechanical stimulation of bone in vivo reduces osteocyte expression of Sost/sclerostin. J. Biol. Chem. **283**(9), 5866–5875 (2008). https://doi.org/10.1074/jbc.M705092200. http://www.ncbi.nlm.nih.gov/pubmed/18089564

47. S. Scheiner, P. Pivonka, C. Hellmich, Coupling systems biology with multiscale mechanics, for computer simulations of bone remodeling. Comput. Methods Appl. Mech. Eng. **254**, 181–196 (2013). https://doi.org/10.1016/j.cma.2012.10.015

48. S. Scheiner, P. Pivonka, D.W. Smith, C.R. Dunstan, C. Hellmich, Mathematical modeling of postmenopausal osteoporosis and its treatment by the anti-catabolic drug denosumab. Int. J. Numer. Method. Biomed. Eng. **30**(August 2013), 1–27 (2014). https://doi.org/10.1002/cnm

49. S. Schmidt, T.M. Post, L.A. Peletier, M.A. Boroujerdi, M. Danhof, Coping with time scales in disease systems analysis: application to bone remodeling. J. Pharmacokinet. Pharmacodyn. **38**(6), 873–900 (2011). https://doi.org/10.1007/s10928-011-9224-2

50. J.M. Spatz, M.N. Wein, J.H. Gooi, Y. Qu, J.L. Garr, S. Liu, K.J. Barry, Y. Uda, F. Lai, C. Dedic, M. Balcells-Camps, H.M. Kronenberg, P. Babij, P.D. Pajevic, The Wnt inhibitor sclerostin is up-regulated by mechanical unloading in osteocytes in vitro. J. Biol. Chem. **290**(27), 16744–16758 (2015). https://doi.org/10.1074/jbc.M114.628313

51. M. Stolina, D. Dwyer, Q.T. Niu, K.S. Villasenor, P. Kurimoto, M. Grisanti, C.Y. Han, M. Liu, X. Li, M.S. Ominsky, H.Z. Ke, P.J. Kostenuik, Temporal changes in systemic and local expression of bone turnover markers during six months of sclerostin antibody administration to ovariectomized rats. Bone **67**, 305–313 (2014). https://doi.org/10.1016/j.bone.2014.07.031

52. C.C. Tang, C. Benson, J. McColm, A. Sipos, B. Mitlak, L.J. Hu, *Population Pharmacokinetics and Pharmacodynamics of Blosozumab*. J. Pharmacokinet. Pharmacodyn. **42**, S80–S81 (2015)

53. S. Taylor, M.S. Ominsky, R. Hu, E. Pacheco, Y.D. He, D.L. Brown, J.I. Aguirre, T.J. Wronski, S. Buntich, C.A. Afshari, I. Pyrah, P. Nioi, R.W. Boyce, Time-dependent cellular and transcriptional changes in the osteoblast lineage associated with sclerostin antibody treatment in ovariectomized rats. Bone **84**, 148–159 (2016). https://doi.org/10.1016/j.bone.2015.12.013

54. S. Trichilo, P. Pivonka, in *Application of Disease System Analysis to Osteoporosis: From Temporal to Spatio-Temporal Assessment of Disease Progression and Intervention* (Springer International Publishing, Cham, 2018), pp. 61–121. https://doi.org/10.1007/978-3-319-58845-2_2

55. S. Trichilo, S. Scheiner, M. Forwood, D.M. Cooper, P. Pivonka, Computational model of the dual action of PTH – application to a rat model of osteoporosis. J. Theor. Biol. **473**, 67–79 (2019). https://doi.org/10.1016/j.jtbi.2019.04.020. https://linkinghub.elsevier.com/retrieve/pii/S0022519319301699

56. K. Warmington, S. Morony, I. Sarosi, J. Gong, P. Stephens, D.G. Winkler, M.K. Sutherland, J.A. Latham, H. Kirby, A. Moore, M. Robinson, P.J. Kostenuik, W.S. Simonet, D.L. Lacey, C. Paszty, Sclerostin antagonism in adult rodents, via monoclonal antibody mediated blockade, increases bone mineral density and implicates sclerostin as a key regulator of bone mass during adulthood. J. Bone Miner. Res. **19**, S56–S57 (2004)

57. R.S. Weinstein, Glucocorticoid-induced osteoporosis and osteonecrosis. Endocrinol. Metab. Clin. North Am. **41**(3), 595–611 (2012). https://doi.org/10.1016/j.ecl.2012.04.004

58. R.S. Weinstein, R.L. Jilka, A. Michael Parfitt, S.C. Manolagas, Inhibition of osteoblastogenesis and promotion of apoptosis of osteoblasts end osteocytes by glucocorticoids potential mechanisms of their deleterious effects on bone. J. Clin. Invest. **102**(2), 274–282 (1998). https://doi.org/10.1172/JCI2799

59. P.C. Witcher, S.E. Miner, D.J. Horan, W.A. Bullock, K.E. Lim, K.S. Kang, A.L. Adaniya, R.D. Ross, G.G. Loots, A.G. Robling, Sclerostin neutralization unleashes the osteoanabolic effects of Dkk1 inhibition. JCI Insight **3**(11) (2018). https://doi.org/10.1172/jci.insight.98673

60. G. Zaman, A.A. Pitsillides, S.C. Rawlinson, R.F. Suswillo, J.R. Mosley, M.Z. Cheng, L.a. Platts, M. Hukkanen, J.M. Polak, L.E. Lanyon, Mechanical strain stimulates nitric oxide production by rapid activation of endothelial nitric oxide synthase in osteocytes. J. Bone Miner. Res. **14**(7), 1123–1131 (1999). https://doi.org/10.1359/jbmr.1999.14.7.1123

Part II
Topics in Patient-Specific Computations

Towards Visualising and Understanding Patient-Specific Biomechanics of Abdominal Aortic Aneurysms

K. R. Beinart, George C. Bourantas, and Karol Miller

Abstract An abdominal aortic aneurysm (AAA) is a permanent and irreversible dilation of the lower aortic region. The current clinical rupture risk indicator for AAA repair is an anterior-posterior AAA diameter exceeding 5.5 cm. This is an inadequate rupture risk indicator given that 60% of AAAs with larger diameters than 5.5 cm often remain stable for the patient's lifetime while 20% of smaller AAAs have ruptured. A more robust predictor of rupture risk is therefore crucial to save lives and reduce medical costs worldwide. Rupture is a local failure of the wall that occurs when local mechanical stress exceeds local wall strength. A comparison of the AAA tension and stretch during the cardiac cycle will provide the indication of wall structural integrity necessary for reliable rupture risk stratification. Employing engineering logic, mismatches between tension and stretch are likely to indicate localized wall weakening and the likelihood of rupture (e.g. a high stretch resulting from a low tension). Biomechanics based Prediction of Aneurysm Rupture Risk (BioPARR) is an AAA analysis software application that currently only determines aneurysm wall tension. This study seeks to investigate the feasibility of determining surface stretches within the AAA wall using methods compatible with clinical practices. It additionally aims to create and validate a new procedure for AAA rupture risk stratification.

Keywords Abdominal aortic aneurysm · Rupture · Computed tomography angiography · Time-resolved · Four-dimensional · Synthetic · Tension · Stretch

1 Introduction

An abdominal aortic aneurysm (AAA) is a permanent and irreversible dilation of the lower aortic region. The condition is usually symptomless and is typically detected during an unrelated procedure. If left untreated, the aneurysm can dissect or rupture

K. R. Beinart · G. C. Bourantas · K. Miller (✉)
Intelligent Systems for Medicine Laboratory, The University of Western Australia, Perth, WA, Australia
e-mail: karol.miller@uwa.edu.au

© Springer Nature Switzerland AG 2020
K. Miller et al. (eds.), *Computational Biomechanics for Medicine*,
https://doi.org/10.1007/978-3-030-42428-2_7

with the high mortality rates of approximately 80–90% [1]. Considering the dangers and expenses related to the surgical treatment, rupture risk classification is essential. If this rupture risk outweighs the risk of surgery, the patient will be considered for endovascular (EVAR) or open repair surgery.

The current clinical rupture risk indicator for repair is an anterior-posterior AAA diameter exceeding 5.5 cm or a diameter growth rate greater than 1 cm/year [2]. This is an inadequate rupture risk indicator given 60% of AAAs with larger diameters than 5.5 cm often remain stable for the patient's lifetime [3] while 20% of smaller AAAs have ruptured [4]. Additionally, AAA rupture has been linked to other risk factors, including: genetic history, smoking, high mean arterial pressure (MAP), gender, vessel asymmetry, growth of intraluminal thrombus (ILT) and increased metabolic activity [5, 6]. Simplistic conclusions based on diameter alone are thus inadequate. A more robust and reliable predictor of rupture risk is therefore crucial to save lives and reduce medical costs worldwide.

Many researchers believe that a patient specific biomechanics-based approach is a promising alternative that could significantly improve the clinical management of AAA patients. With recent advancements in medical imaging and analysis software, geometrically accurate patient specific AAA three-dimensional (3D) models can now be constructed for the purpose of computer simulations that calculate wall stress. Studies have demonstrated that peak wall stress is a better indicator of individual rupture risk compared to aortic diameter [7]. Stress alone, however, will not provide an accurate estimation of rupture risk as mechanical failure of the wall is dependent on both local wall stress and local wall strength. Vande Geest et al. derived a statistical model for the non-invasive estimation of wall strength [8]. This strength model, however, is population-based, not patient specific and moreover not localized.

Many studies have utilized displacement tracking algorithms on time-resolved (4D) ultrasound scans to investigate local AAA wall deformations [9]. High local strains alone, however, cannot provide an indication of wall strength, as they may be generated by high local wall tensions.

AAA rupture is a local failure of the wall that occurs when local mechanical stress exceeds local wall strength [10]. This study proposes that a comparison of AAA tension with stretch during the cardiac cycle will provide the indication of wall structural integrity necessary for reliable rupture risk stratification. It is hypothesized that mismatches between local tension and resulting tangential stretch, such as high stretch with low tension, indicate localised wall weakening and the likelihood of rupture.

Biomechanics based Prediction of Aneurysm Rupture Risk (BioPARR) is an existing, free and semi-automatic AAA analysis software application that currently only determines aneurysm wall tension [11]. This study seeks to investigate the feasibility of determining surface stretches within the AAA wall using methods compatible with clinical practice. It additionally aims to validate the approach of pairing surface stretches with tension as a measure of AAA rupture potential.

A variety of approaches have been utilized by researchers to obtain ground truth data for validation purposes. Most methods are inaccurate and inefficient due to the errors introduced by reference tracking algorithms, sparse location of reference markers and the bias introduced by these markers on the tracking problem. Additionally, fabrication of physical phantoms to simulate realistic physiological deformation is both challenging and expensive.

Synthetic data provides a valuable reference for assessing the accuracy of tracking algorithms due to knowledge of the exact deformation. In this case, the reference displacement field is unbiased by any motion estimation algorithm. Additionally, exact deformation is known at each voxel. Furthermore, a wide range of digital data can easily be created by researchers thus eliminating the requirement for complex experimental phantoms. The usefulness of synthetic data as a validation tool, however, is highly dependent on the degree of realism of the generated synthetic scans.

One method of creating synthetic datasets involves the use of algorithms that simulate the physics of the imaging process. Models of virtual patient anatomy can consequently be 'imaged' using these projection algorithms. Models of the patient anatomy are only simplified geometries that have been mathematically derived and are therefore largely unrealistic. Furthermore, the organs and substructures are modelled as homogenous with constant pixel intensity. Image artefacts introduced by the heterogenous tissues are not simulated [12]. Therefore, although these phantoms can be used for dosimetry studies, they are inadequate for reliably assessing techniques dependent on image quality.

In the pursuit of increasingly realistic synthetic data, new techniques use biomechanical models extracted from the segmentation of real patient anatomy. A single static real medical scan is then warped with the deformation field of this model [13]. The use of real scans enables more accurate synthetic data creation by accounting for the heterogeneous tissue voxel intensities. Exact and simple methods to achieve this have not been clearly outlined in the literature. Additionally, these methods have mainly been restricted to the modelling of cardiac motion using only echocardiography and MRI [13]. This study therefore additionally aims to extend the existing literature by developing and clearly outlining simple methods for the simulation of realistic CT images using open source software for the given application of AAA.

2 Methods

2.1 Synthetic Data

A simple method of creating a synthetic 4D CT dataset was developed. This was achieved by warping a static CT scan using the transformation matrices obtained after modelling the pulsatile motion of the abdominal aortic aneurysm geometry.

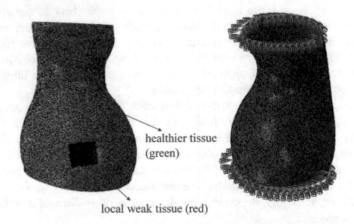

Fig. 1 Left: The local weakened (red) and healthier (green) tissue regions of the model. Right: Fixed Boundary Conditions applied to the ends of the AAA model

One abdominal aortic aneurysm computed tomography angiography DICOM scan was provided by Dr. Hozan Mufty of UZ Leuven academic hospital, Belgium. A 3D model of the AAA was created by segmenting the CT scan in 3D Slicer 4.10.1, a free open source medical image analysis and visualization software package.

The outer wall of the abdominal aortic aneurysm model was imported into Abaqus Explicit 2018. This was taken as the geometry that had been pre-loaded with the diastolic pressure. A linear tetrahedral element mesh was used due to its compatibility with Abaqus Explicit. The mesh contained approximately 4×10^6 nodes. The simulation consisted of a periodic loading cycle using an internal pulsatile pressure of 10 kPa. This represents a high pulse pressure that would realistically be observed in AAA patients. The upper and lower ends of the aneurysm were constrained in all directions using fixed boundary conditions (Fig. 1). Non-linear, hyper-elastic material properties were used to model the aneurysm tissue using the strain energy function presented by Raghavan and Vorp [14]. This strain energy function (W) shown below, was obtained by the researchers after examining the mechanical properties of excised AAA tissue.

$$W = a\,(I_{1c} - 3) + b(I_{1c} - 3)^2 \tag{1}$$

a and b are the material properties and I_{1c} is the first invariant of the right Cauchy-Green tensor. Most of the aneurysm tissue was modelled using $a = 113.4$ kPa, $b = 9.2$ kPa and a density of 1000 kg/m^3 [15]. A randomly chosen local region of the aneurysm model was purposely weakened by halving each of these material parameters. In addition to location, the extent and range of weakening was arbitrarily selected. The local weakened and healthier tissue regions are indicated in Fig. 1 in red and green respectively.

Mesh nodal coordinates from five phases of the pulsating biomechanical model, between the two extremes of 'diastole' and 'systole', were extracted and exported from Abaqus to 3D Slicer. The transformation matrices, mapping each of the nodal coordinates from phase 0 to each of the respective phases, were obtained using the 'Scattered Transform' module [16]. The module interpolates displacements at nodes using a BSpline Algorithm. Once the transformation matrices were obtained, the 4D synthetic dataset was created using the 'Data' module. The initial CT scan was warped by each of these transformation matrices after dragging and dropping it onto the relevant transform. The new CT frames were then saved by hardening the transforms onto the volume. This resulted in a stack of synthetic CTs corresponding to each phase of the pulsating biomechanical model.

2.2 Voxel Displacement Tracking

As an alternative to producing an in-house code for the implementation of the displacement tracking techniques, open-source tools are available, such as those used for the registration of medical scans. Registration is the task of mapping one image to another image. This is typically used by clinicians to align scans of different modalities, or even align scans taken at different points in time such as for follow up procedures. Registration can therefore also be used to determine displacements of the aneurysm wall from scans at different points in time during the cardiac cycle.

Thirion proposed the Demons algorithm for non-rigid registration [17]. The Diffeomorphic Demons algorithm minimizes the sum of square differences of intensity, contains a smoothness constraint and additionally limits the transformation to be one-to-one. The Demons algorithm embodies a computationally efficient simplification of the optical flow problem.

The Demons Diffeomorphic Registration was implemented in 3D Slicer using the 'BRAINSDemonWarp' module. A course-to-fine pyramidal approach was utilized using 5 pyramid levels. A shrink factor of 16 and iteration count of 300, 50, 30, 20 and 15 for each respective pyramid level was employed. Linear interpolation and a Diffeomorphic Registration Filter were used. These parameter settings produced the most accurate results when visually compared with ground truth.

Each synthetic CT frame was registered to the initial frame. The outputs of these registrations were transformation matrices mapping points from one image to the next. The transformation matrices were then converted to displacement fields in the 'Transforms' module. Using the 'Probe Volume' module, the displacement field was then overlayed onto the surface of the segmented aneurysm geometry.

2.3 Determining Maximum Principal Stretch

The point coordinates of the AAA surface and the displacements at these nodes were read into MATLAB. An in-house modified moving least squares (MMLS) code was utilized in order to determine the deformation gradient from these nodal displacements [18]. The deformation gradient (F) was obtained by determining the derivative of the displacement vectors with respect to the undeformed configuration (X) and adding the identity matrix (I):

$$F = I + \frac{\partial u}{\partial X} \tag{2}$$

Additional code was added in order to determine the principal stretches. We computed the right Cauchy Green strain tensor: $C=F^T F$. Eigenvalues of the right Cauchy Green strain tensor are the square of the principal stretches. The maximum principal tangential stretches and its directions were obtained after aligning the minimum eigenvectors with the surface normals. This is compatible with reality whereby the aorta wall will compress radially but stretch tangentially when it is inflated by the blood pressure.

2.4 Determining Maximum Principal Tension

The Maximum Principal Tension was determined via BioPARR utilizing the following inputs: a constant wall thickness of 1 mm, 16 kPa pressure applied to the interior AAA surface representing the patient's mean arterial blood pressure and a ten-node tetrahedral hybrid element (C3D10H) mesh. The 'no ILTP' case was modelled. This case ignores the intraluminal thrombus and loads the interior surface of the AAA with blood pressure. This was done for simplicity and because the ILT was neglected when modelling the AAA motion.

2.5 New Rupture Risk Index

The MATLAB code was additionally updated to read-in the maximum principal tensions obtained from BioPARR. A structural integrity index (SII) was created by dividing the maximum principal tension by the largest maximum principal stretch during the cardiac cycle. A relative structural integrity index map (RSII) was created by dividing the SII map by the maximum structural integrity index over the AAA volume. This enables clear visualization of weakened areas by comparing all the structural integrity indices over the AAA volume with the strongest tissue present.

2.6 Validation of Techniques

The technique was validated by correlating displacements and maximum principal stretches obtained from 4D CT registration with the ground truth values obtained from Abaqus. This was implemented for each phase of the cardiac cycle. A Pearson correlation test was conducted in Excel with significance evaluated using a p-value of 0.05. Similarity to ground truth was also observed by visualizing displacements and maximum principal stretches in Paraview, an open-source data analysis and visualization application.

This new rupture risk predictor was then validated by determining if the randomly located purposely weakened area of the model was detected. This was achieved by visualizing relative structural integrity indices below 0.15 using Paraview. This represents the weakest 15% of tissue within the AAA.

3 Results

3.1 Validation of Displacement Tracking

A high similarity was observed between the ground truth displacement fields obtained via Abaqus and that obtained from registration of the synthetic 4D CT scans. This is depicted in Fig. 2 which displays the tangential displacement fields of the abdominal aortic aneurysm model during one phase of the cardiac cycle. This is additionally indicated by the high Pearson's correlation coefficients of displacement

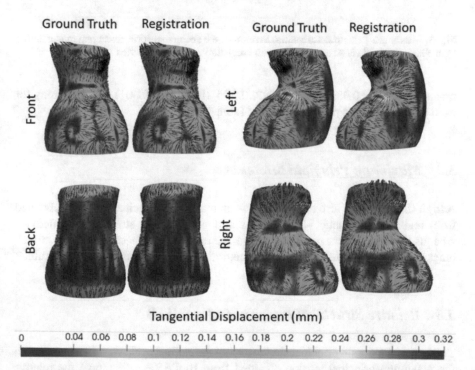

Fig. 2 Tangential displacements of the abdominal aortic aneurysm model during one phase of the cardiac cycle

Table 1 Correlation coefficients for each phase of the cardiac cycle

Frame	Correlation (X)	Correlation (Y)	Correlation (Z)	Correlation (magnitude)	P-value
1	0.99961	0.99930	0.99575	0.98571	P<0.001
2	0.98952	0.99934	0.99628	0.98952	P<0.001
3	0.99971	0.99952	0.99674	0.99347	P<0.001
4	0.99976	0.99966	0.99722	0.99602	P<0.001
5	0.99975	0.99965	0.99684	0.99750	P<0.001

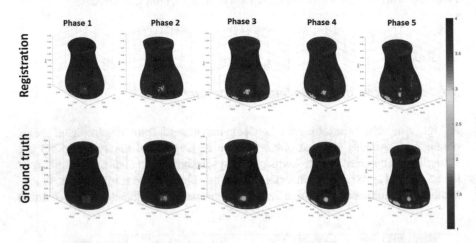

Fig. 3 Maximum Principal Tangential Stretch of the abdominal aortic aneurysm model during each phase, obtained via Abaqus (bottom) and registration of 4D synthetic CT scans (top)

magnitudes (R = 0.986, 0.990, 0.993, 0.996, 0.998, p < 0.001) and directions for each of the respective phases analysed (Table 1).

3.2 Maximum Principal Stretches

A high similarity was also observed between maximum principal stretches obtained from registered synthetic 4D CT scans and ground truth stretches obtained via Abaqus. This is evident in Fig. 3, where for each of the phases analyzed, stretch magnitudes and patterns obtained via registration are comparable to ground truth.

3.3 Relative Structural Integrity Index (RSII)

The largest maximum principal stretch during the cardiac cycle was then paired with the maximum principal tension obtained from BioPARR to compute the relative structural integrity index (RSII). A correlation analysis between the ground truth and

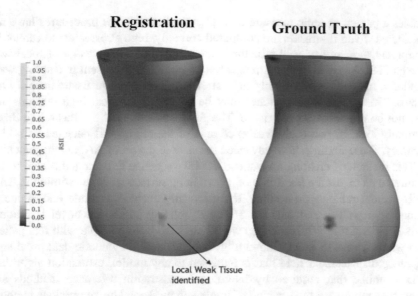

Fig. 4 Lowest 15% of relative structural integrity indices (RSII) of the aneurysm model

registered RSII distributions indicated that good agreement was obtained ($R = 0.98$, Pearson's correlation, $p < 0.001$). As evident in Fig. 4, an illustration of the lowest 15% of RSII successfully identifies the purposely locally weakened tissue depicted in Fig. 1.

4 Discussion

This study has successfully developed a procedure to accurately determine surface stretches within the AAA wall using methods compatible with clinical practices.

Most researchers have focused on utilizing time-resolved ultrasound to determine deformation of AAAs. This study has highlighted the feasibility of using 4D CT as an alternative. This is compatible with clinical workflow due to the current practice of employing 3D CT angiography for preoperative imaging of the AAA. Unlike ultrasound, 4D CT additionally enables quick, repeatable acquisition of the full volume of the AAA.

The use of the Demons Diffeomorphic registration technique to track deformation during the cardiac cycle from 4D CT scans was validated. The obtained displacements and resulting stretches were highly accurate with strong correlation to ground truth.

This novel study has introduced a new and improved rupture risk metric. The RSII utilizes a holistic engineering approach by accounting for both local stretches and tensions to enable the characterization of tissue integrity local to the AAA. This enables a patient specific measure of wall strength that other procedures have not considered. Even if stresses are computed correctly, high stresses alone cannot be interpreted as a loss of wall structural integrity without knowledge of local wall strength. i.e. clearly high wall stress is not an issue if it is present in a strong wall. Similarly, methods utilizing only high stretch as a measure of tissue integrity are flawed. These local high stretches may be generated by local high tensions and may not be due to weakened tissue. The RSII was validated by illustrating that a randomly located, purposely weakened area of the model was detected with high accuracy. These findings have advanced the state of the art of AAA management.

This method of creating a synthetic 4D CT sequence has granted access to the required data to test the feasibility of determining surface stretches within the AAA wall, without reliance on a clinic. It additionally enabled accurate knowledge of ground truth values and thus the ability to reliably assess the novel techniques used. This essential validation step would not have been possible with real patient data where access to exact ground truth is unattainable. Synthetic data provides a reference displacement field that is unbiased to any motion estimation algorithm. This is unlike that required by intermodal registration reference methods and techniques relying on the tracking of implanted markers. Unlike previous methods that utilise sparsely located reference markers, the technique used in this study provides knowledge of exact deformation at each voxel. Furthermore, the simple, low cost computer-based biomechanical model is more realistic compared to other mock-ups such as complicated physical phantoms, due to easier control of material properties and pressures. This opens the door to the generation of a wide range of synthetic data, from normal to varying diseased states, as demonstrated by this AAA study. The usefulness of synthetic data as a validation tool, however, is highly dependent on the degree of realism of the generated sequence. Unlike synthetic datasets created using projection algorithms, this study uses methods that produce realistic synthetic data. This was achieved by using real scans to extract exact patient anatomy and to simulate the heterogenous voxel intensities of imaged tissue.

The simple and easily accessible methods developed in this study can similarly be used by other researchers to progress pilot studies without being impeded by clinical bureaucracy. Additionally, the flexibility offered by this simple technique provides a platform to optimize and validate emerging technologies and methods without being impeded by the multitude of external restrictions imposed by the other validation techniques discussed.

Limitations, however, do exist in the presented work. This method of synthetic CT creation does not completely take the physics of image acquisition into account. Instead it re-uses the same texture of the initial CT, which is warped according to the differences between the original scan and the simulated motion. Changes in the geometry of the moving organ, however, will alter the path length along which the radiation travels through the organ. This will cause variations in voxel intensity throughout the cardiac cycle. The change in voxel intensity during deformation is not reflected in the synthetic data creation technique discussed.

One method discussed in the literature partly accounts for this by using a template 4D DICOM dataset to partially increase the degree of realism of the generated synthetic sequence [19]. This is achieved by spatio-temporal alignment of the template sequence with the biomechanical model. In this method instead of warping a single static scan at the initial phase of the cardiac cycle, the template scan is warped by the biomechanical model at each of the respective phases. This partially accounts for the change in intensities that will be present as a result of deformation because it reduces the difference between the reference and deformed frames. The risk of unrealistic texture warping does, however, still exist with this method when the simulated motion of the model deviates too far from the template motion. That method, however, requires the presence of an initial 4D dataset. In novel studies such as this one, access to an initial 4D dataset is not always possible. A 4D CT protocol of the AAA is not yet utilized in the clinic. Once access to real data from this protocol is achieved, a future study can further validate the methods used by implementing this improved technique.

A basic assumption made using the Demons algorithm is that the intensity of voxels remains constant through time. The geometry of the aneurysm, however, will be changing during the cardiac cycle, which, as discussed, will alter voxel intensities. Since this synthetic data is slightly unrealistic in that the intensity of voxels remains constant despite motion, the methods used on this artificial dataset are acceptable. When using real data, however, this may not remain true. An option for dealing with this issue could be to not register each frame to the initial frame, as was done using this synthetic dataset. Instead one could register each frame to the previous frame but use the preceding transform as an initialization to the registration. This would enable the constant intensity assumption to hold true as the geometry between consecutive frames would not change significantly.

The next step required to progress this novel technique into normal clinical practice is an initial pilot study using real patient data. Further studies will need to establish the relationship between RSII and the progression of abdominal aortic aneurysms using follow up analyses.

Acknowledgements The authors wish to acknowledge help from is research was supported by Dr. Grand Roman Joldes, Dr. Mark Teh (Director of Training, Radiology, SCGH), Dr. Benjamin Khoo (Medical Physicist, SCGH), Dr. Toby Richards (HoD of Vascular Surgery, Fiona Stanley Hospital), Dr. Hozan Mufty (Vascular Surgeon, UZ Leuven academic hospital) and Dr. Abdul Maher (Vascular Surgeon, SCGH).

References

1. L.L. Hoornweg et al., Meta analysis on mortality of ruptured abdominal aortic aneurysms. Eur. J. Vasc. Endovasc. Surg. **35**(5), 558–570 (2008)
2. B.J. Doyle et al., On the influence of patient-specific material properties in computational simulations: A case study of a large ruptured abdominal aortic aneurysm. Int. J. Numer. Methods Biomed. Eng. **29**(2), 150–164 (2013)
3. D. Farotto et al., The role of biomechanics in aortic aneurysm management: Requirements, open problems and future prospects. J. Mech. Behav. Biomed. Mater. **77**, 295–307 (2018)

4. S.C. Nicholls et al., Rupture in small abdominal aortic aneurysms. J. Vasc. Surg. **28**(5), 884–888 (1998)
5. N. Sakalihasan et al., Positron emission tomography (PET) evaluation of abdominal aortic aneurysm (AAA). Eur. J. Vasc. Endovasc. Surg. **23**(5), 431–436 (2002)
6. T.C. Gasser et al., A novel strategy to translate the biomechanical rupture risk of abdominal aortic aneurysms to their equivalent diameter risk: Method and retrospective validation. Eur. J. Vasc. Endovasc. Surg. **47**(3), 288–295 (2014)
7. D.A. Vorp, Biomechanics of abdominal aortic aneurysm. J. Biomech. **40**(9), 1887–1902 (2007)
8. J. Vande Geest et al., Towards a noninvasive method for determination of patient-specific wall strength distribution in abdominal aortic aneurysms. Ann. Biomed. Eng. **34**(7), 1098–1106 (2006)
9. K. Karatolios et al., Method for aortic wall strain measurement with three-dimensional ultrasound speckle tracking and fitted finite element analysis. Ann. Thorac. Surg. **96**(5), 1664–1671 (2013)
10. T.C. Gasser, Biomechanical rupture risk assessment: A consistent and objective decision-making tool for abdominal aortic aneurysm patients. Aorta J. **4**(2), 42–60 (2016)
11. G. Joldes et al., BioPARR: A software system for estimating the rupture potential index for abdominal aortic aneurysms. Sci. Rep. **7**(1), 4641–4641 (2017)
12. N. Lowther et al., Investigation of the XCAT phantom as a validation tool in cardiac MRI tracking algorithms. Phys. Med. **45**, 44–51 (2018)
13. P. Clarysse et al., *Simulation based evaluation of cardiac motion estimation methods in tagged-MR image sequences.* J. Cardiovasc. Magn. Reson. **13**(Suppl 1), P360 (2011)
14. M.L. Raghavan, D.A. Vorp, Toward a biomechanical tool to evaluate rupture potential of abdominal aortic aneurysm: Identification of a finite strain constitutive model and evaluation of its applicability. J. Biomech. **33**(4), 475–482 (2000)
15. A. Satriano et al., In vivo strain assessment of the abdominal aortic aneurysm. J. Biomech. **48**(2), 354–360 (2015)
16. G.R. Joldes et al., *Performing Brain Image Warping Using the Deformation Field Predicted by a Biomechanical Model* (Springer, New York, 2012), pp. 89–96
17. J.P. Thirion, Image matching as a diffusion process: An analogy with Maxwell's demons. Med. Image Anal. **2**(3), 243–260 (1998)
18. G.R. Joldes et al., Modified moving least squares with polynomial bases for scattered data approximation. Appl. Math. Comput. **266**(C), 893–902 (2015)
19. N. Duchateau et al., Model-based generation of large databases of cardiac images: Synthesis of pathological cine MR sequences from real healthy cases. IEEE Trans. Med. Imaging **37**(3), 755–766 (2018)

Pipeline for 3D Reconstruction of Lung Surfaces Using Intrinsic Features Under Pressure-Controlled Ventilation

Samuel Richardson, Thiranja P. Babarenda Gamage, Toby Jackson,
Amir HajiRassouliha, Alys Clark, Martyn P. Nash, Andrew Taberner,
Merryn H. Tawhai, and Poul M. F. Nielsen

Abstract The measurement of whole lung mechanics forms the basis of diagnostic measurements for many respiratory diseases. Despite this, there are currently no quantitative methods to link alterations in pulmonary microstructures to measurements of whole lung function. The normal decline in the lung's microstructure that occurs with age is virtually indistinguishable from early disease on imaging or standard lung function measurements, leading to frequent misdiagnosis in the elderly. Accurate characterisation of lung mechanics across spatial scales has the potential to assist distinguishing age from pathology, which would benefit patients across a range of medical conditions and procedures. While computational modelling promises to be a useful tool for improving our understanding of lung mechanics, there is currently no unified structure-function computational model that explains how age-dependent structural changes translate to decline in whole lung function. This paper presents novel instrumentation and imaging techniques for measurements of intact *ex vivo* lung tissue mechanics. We seek to address problems of weak parameterisation that existing models suffer from, due to lack of reliable measurements. To begin addressing this issue, we have developed a full-field stereoscopic imaging system for tracking surface deformation of the rat lung during pressure-controlled ventilation. This study presents a pipeline for the reconstruction and tracking of the intact left lobe of a rat lung during inflation, *ex vivo*. Model-based 3D reconstruction of the lungs enabled the 3D shape of a surface patch of the imaged lung to be determined. The 3D reconstruction and tracking of the fresh lung surface patch in this study was completed with three cameras across 21 pressure steps, encompassing a total pressure change from 2069 Pa to

S. Richardson · T. P. Babarenda Gamage · T. Jackson · A. HajiRassouliha · A. Clark · M. H. Tawhai
Auckland Bioengineering Institute, University of Auckland, Auckland, New Zealand

M. P. Nash · A. Taberner · P. M. Nielsen (✉)
Department of Engineering Science, Auckland Bioengineering Institute, University of Auckland, Auckland, New Zealand
e-mail: p.nielsen@auckland.ac.nz

© Springer Nature Switzerland AG 2020
K. Miller et al. (eds.), *Computational Biomechanics for Medicine*,
https://doi.org/10.1007/978-3-030-42428-2_8

123

2386 Pa. This approach shows that reconstructing intact *ex vivo* fresh lungs, with no additional surface markers, is feasible.

Keywords Image reconstruction · Lung · Respiratory disease · Computational modelling · Lung tissue mechanics · Pulmonary microstructure

1 Introduction

Despite the importance of the lungs in delivering oxygen to the body, aspects of their mechanics remain poorly understood [1]. A key reason for this is that any disruption of the lung structure results in a change in the mechanical response of the tissue, making traditional mechanical testing poorly suited to investigating lung tissue [2]. Many studies have attempted to characterise the mechanics of lung tissues, however, it was not until the middle-to-late twentieth century that respiratory mechanics began to be studied as a separate field, and it was during this time that the majority of our understanding was developed [3, 4]. Despite advances in imaging technologies, fundamental questions concerning key processes that occur in the lungs remain unanswered. For example, there is no unifying theory for alveolar dynamics and recruitment during respiration. It remains unclear if the alveoli expand isotropically, heterogeneously, or by a combination of both [5]. This has been debated in the literature, with consensus being hindered by difficulties in imaging the small and constantly moving alveoli during respiration.

Computational modelling may prove to be a useful tool for improving our understanding of lung mechanics, and several computational models have been proposed for the mechanics of lung tissue. However, there is currently no unified structure-function computational model that explains how age-dependent structural changes translate to decline in whole lung function. Existing models suffer from weak parameterisation due to lack of available data. In this study, we designed a real-time full field stereoscopic imaging system for tracking lung surface deformation under pressure-controlled inflation. This system will enable us to acquire rich, accurate, robust, and previously unavailable physiological data on lung tissue mechanics from whole rat lungs, that can ultimately be used to parameterise computational models of lung mechanics.

2 Methodology

2.1 Lung Ventilation

Fresh post-mortem lungs were acquired from female (350 ± 50) g Sprague-Dawley rats, after the animals were sacrificed following separate experimental studies that did not involve the chest cavity. The Sprague-Dawley strain was chosen for two key reasons: similarities to humans in alveolar air-space enlargement with age [6]; and their relatively large alveoli (~90 μm diameter) [6] compared with lung size (~20 mL). A cannulated rat lung is shown in Fig. 1.

Fig. 1 Inflated left lung lobe
held at 3000 Pa, in a Petri
dish full of phosphate
buffered saline solution and
cannulated with a plastic 16
Gauge blunted needle

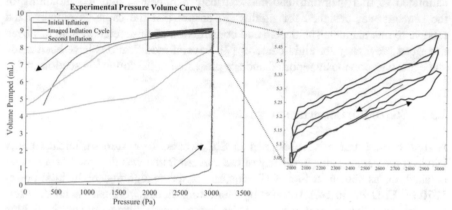

Fig. 2 Left, PV loops from two full range inflations and an imaging cycle of three PV loops from
2000 Pa to 3000 Pa and back. Arrows depict the direction of increasing time. Right, expanded view
of the three PV loops used for imaging

A CompactRio (National Instruments) based real time pressure control system
was developed to control the inflation of the lungs. A syringe pump enabled real
time pressure control, with volume and pressure resolutions of ±5 μl and ± 5 Pa
respectively. A 100 ml glass syringe was mounted and actuated by a Physik
Instrumente DC-Mike linear actuator that has an encoder resolution of 0.0592 μm.

During stereoscopic imaging of the lungs, images were captured at regular
intervals corresponding to increments/decrements in pressure of 15 Pa. Figure 2
shows the pressure-volume (PV) loops from the stereoscopic measurement of the
lung lobe. The imaged inflation cycle (red in Fig. 2) shows three cycles between
2000 Pa and 3000 Pa. The PV loops between 2000 Pa and 3000 Pa are approximately
linear, with a small amount of hysteresis visible between 2800 Pa and 3000 Pa.
There was an increase in lung volume of 0.2 mL across the three loops, when
comparing the volumes at 2000 Pa.

2.2 Lung Surface Imaging

A 12 camera full field stereoscope was designed and built in-house to enable imaging of the surface displacement of the lung during pressure-controlled inflation. FLIR BlackflyS monochrome cameras that feature a SONY IMX250 sensor were selected for imaging the lung due to their high quantum efficiency and high signal to noise ratio (4760 signal to noise ratio or 73 dB dynamic range). The sensors had a 2448 pixel × 2048 pixel resolution (5.0 MP) with a 3.45 μm pixel size and were capable of imaging at 75 frames per second. The control code for these cameras was written in LabView (National Instruments), enabling data from all 12 cameras to be saved concurrently.

To ensure accurate 3D reconstruction of the imaged objects, the cameras were calibrated to find their intrinsic and extrinsic parameters, and the mounting of the cameras was designed for rigidity, to ensure that the cameras remain fixed relative to one another. The design and construction of this stereo system has been described previously for eight cameras [7]. Several modifications have been made since this was previously reported and are presented in the following sections.

2.2.1 Stereo Rig Construction

A rigid camera frame was designed in Solidworks. To ensure sufficient rigidity between the cameras, the geometry of the camera frame was designed as a regular octahedron, as shown in Fig. 3. To ensure consistent lighting, eight high-power 1270 lm LED Engin LZ1-10R200 light emitting diodes were used with diffusers to ensure even lighting and to reduce noise in the camera images. Image acquisition from the cameras was performed in LabVIEW and the cameras were synchronized using a hardware trigger from the pressure control FPGA. This enabled images to be triggered, based on changes in pressure.

Lungs were dissected from the rats *en bloc*, with the heart and trachea attached. The heart and right lobes were removed, leaving the left lobe and a length of bronchus for cannulation. After cannulation of the lungs onto a blunted needle, they were attached to the syringe pump system. This enabled the initial inflation of the lungs from their collapsed state. The lungs were inflated to a pressure of 3000 Pa and held at that pressure until fully inflated. After a full inflation/deflation cycle, the lungs were bathed in phosphate buffered saline (PBS) to ensure that they remained hydrated. Post hydration, the lungs were mounted into the centre of the stereo camera system.

2.2.2 Stereo Rig Calibration

Camera calibration is necessary to achieve high accuracy imaging and 3D reconstruction. The accuracy of any 3D measurement made with a stereo imaging system

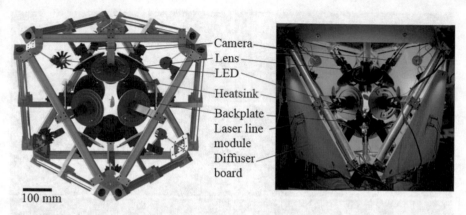

Fig. 3 Frame constructed for performing full-field imaging of the lung surface during pressure-controlled inflation experiments. Left shows a CAD rendering of the stereo rig, Right shows the physical rig

depends, in part, on the accuracy of the calibration of the stereo cameras. The process of calibrating a camera system is a complex problem, which grows in complexity with every additional camera. The calibration method used in this study was developed by HajiRassouliha et al. [8] using a checkerboard calibration template. This has been described by HajiRassouliha et al. in [8] for cameras where all cameras could see the same calibration template. In this study, we extended the calibration approach to allow for calibration of all cameras in the stereo rig. This involved calibrating overlapping groups of four cameras, followed by an alignment of the calibrated cameras sets using a 3D triangular template with three white cellulose precision microspheres of a known diameter attached to each of its vertices. The diameters and spacing between spheres were identified using micro-CT imaging with a resolution of 2.7 μm.

2.2.3 Initial Surface Reconstruction

The first step in an inflation was to acquire images of the *ex vivo* lung while it was illuminated by a laser line, as depicted in Fig. 4. Images including laser lines were acquired without LED illumination These data were used to generate an initial 3D reconstruction of the lung shape. This involved segmenting and fitting the laser lines on the lung lobe using piecewise cubic splines in each of the 2D images from each camera view. The pixel coordinates of these splines were triangulated into 3D space by determining their locations across multiple cameras using an intersecting ray approach, as described in [9], with the requirement that four rays intersect for a point to be considered valid. This resulted in a 3D point cloud which described the surface of the lung.

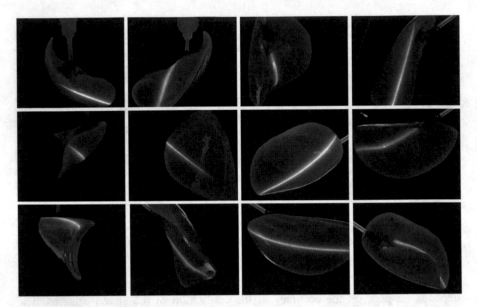

Fig. 4 Examples of laser line images. The lungs were held at a fixed pressure while each line was acquired individually. In this data set, the lungs were held at 2000 Pa. Firstly, images of the lungs were taken at different levels of illuminations from LEDs, then 22 images were recorded of individual laser lines on the lungs

Immediately after laser line data acquisition, the lungs were cyclically inflated and deflated for imaging.

2.3 Lung Fixing and Micro Computed Tomography (CT) Imaging

To obtain an initial estimate of the 3D shape of the lungs, after stereoscopic imaging, lungs were fixed and imaged using a Bruker SkyScan 1272, micro-CT scanner at a pixel resolution of 25 μm. The lungs were fixed by inflating the lungs with 2.5% glutaraldehyde buffered with phosphate buffered saline solution, up to a pressure of 2450 Pa (25 cmH$_2$O). Tissue samples fixed in glutaraldehyde are extensively cross-linked, providing excellent ultrastructural stiffening that maintains the structure of the alveoli, enabling imaging with micro-CT [10]. This process was carried out after stereoscopic imaging, as cross-linking reactions of glutaraldehyde are largely irreversible [11].

The lungs were held at the fixation pressure for 24 h. After 24 h the lungs were attached to a regulated air source, which maintained an even pressure of 2450 Pa (25 cmH$_2$O) to air dry the fixed lungs. The result of this process was a dried lung

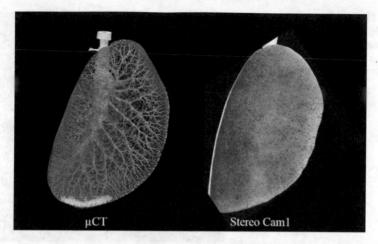

Fig. 5 Left, Micro-CT of the fixed lung lobe. Right, view of fixed speckled lung from a single camera

lobe, with no living tissues, and with the structural proteins cross-linked to maintain the lung structures. An example of this can be seen in Fig. 5.

The micro-CT image of the lung lobe, shown in Fig. 5, enabled the creation of a mesh of the lung lobe. This process started with thresholding of the 2D images to create binary masks. Any holes in the masks were corrected manually. An ITK-based marching cubes algorithm was then implemented to convert each binary mask into a 3D isosurface, which was converted into a point cloud that represented the surface of the lungs from the micro-CT data. While some discrepancies were introduced by the cross-linking procedure and shrinkage during the air-drying process, the mesh of the fixed lung generated from micro CT imaging provided a close approximation to the shape of the unfixed lung.

2.4 Improving Lung Surface Reconstruction and Tracking Motion

The dense point cloud created from the segmented micro-CT data described in Sect. 2.3 was aligned to the sparsely reconstructed laser line data acquired from the stereo rig described in Sect. 2.2.3 using a coherent point drift algorithm to rigidly translate, rotate, and scale the point cloud.

A quadratic Lagrange surface mesh was fitted to the aligned micro-CT point cloud using the fitting algorithms in GIAS2 [12], which minimises the weighted sum of the projections of the point cloud onto the surface. The result of this procedure was an initial surface mesh that was aligned with the position of the stereo-imaged lung, as shown in Fig. 6.

Fig. 6 Fresh lung meshes.
Laser line points are white.
The micro-CT point cloud is
green, and the quadratic patch
Lagrange patch is gold to
black

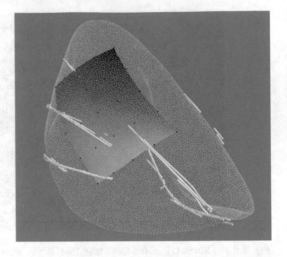

A model-based reconstruction approach was then used to improve upon the initial reconstruction, by mapping texture information across camera views to generate a dense set of corresponding 3D points on the lung surface [9]. In this case, the micro-CT surface mesh was used as a prior model to aid reconstruction of the lung surface. This involved projecting pixels from a reference camera (in this case, Camera 1) onto the quadratic Lagrange micro CT surface mesh. These points were then backprojected to another camera's sensor (in this case, Camera 2) and resampled to generate a new image, which closely resembled the real view from Camera 2. Cross-correlation techniques were then used to identify corresponding points between the resampled image and the real image from Camera 2. These corresponding points were then triangulated to generate a 3D reconstruction of the surface. This operation requires knowledge of the positions of the cameras, which were found during the camera calibration procedure.

The lung surface was reconstructed in this manner at the same inflation pressure used for fixing the lung. The motion of the lung surface during subsequent inflation pressure steps was tracked by performing 2D cross-correlation of the reconstructed corresponding points across the images acquired from each individual camera. These tracked image points were then triangulated to provide a 3D surface reconstruction at each of the inflation pressures.

3   Results

3.1   Tracking of Intrinsic Features

One of the primary concerns with reconstructing and tracking the motion of the fresh lung lobes was the lack of surface texture. To test the ability of the 2D subpixel image registration code [13] to track the intrinsic features of the fresh lung lobe, tracking was performed on a single camera view of a lung across several pressure steps, as shown in Fig. 7.

Confidence thresholds [13] were set to remove points that did not have a strong correlation peak. Figure 7 illustrates that the subpixel image registration method is capable of tracking intrinsic features on the surface of the fresh lung. Failure of the 2D subpixel image registration algorithm would result in no or randomly oriented vectors being returned. The patchy, non-uniform pattern visible in Fig. 7 is a result of the single camera tracking not having sufficient data to capture the displacements of the complex 3D surface of the lung.

3.2   3D Reconstruction Results

To test that reconstruction was effective on fresh lung, a region of interest (ROI) on the back of the lung, which had few specular reflections, was selected, as can be seen in Fig. 8.

Fig. 7 Single camera tracking of the intrinsic features of a left lung lobe. The pressure difference between the reference image and tracked image is shown in the top left

Fig. 8 Region of interest for a reference camera selected on fresh lung. In the reference state the lung was inflated to 2069 Pa

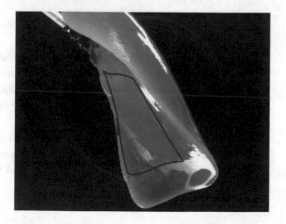

Distance Moved in mm

Fig. 9 Reconstructed lung surface points displayed as spheres, coloured by displacement magnitude, viewed from three angles to display the surface curvature

The model-based reconstruction approach described in Sect. 2.4 was then applied to determine corresponding points with the region of interest across the other cameras in the rig that could see the same region. For the selected ROI, two other cameras could see the same region. The resulting set of corresponding points were then triangulated to find their 3D locations, as seen in Fig. 9.

The 3D locations of these points were then tracked across a range of inflation pressures. This resulted in a 3D deformation field, such as that seen in Figs. 9 and 10.

The 3D reconstruction of the fresh lung enabled tracking of the motion of the lung as a result of pressure increases. In this study, the fresh lung was tracked across a pressure change of 317 Pa. Over this range, the mean magnitude of the 3D motion (0.525 mm) was computed by determining the Euclidean distances between point positions at each pressure. Areas of non-uniformities in the displacement vectors are likely due failure to identify corresponding points across the three cameras. Spurious vectors could be eliminated by adjusting the cross-correlation confidence thresholds to be appropriate for 3D tracking.

4   Summary

This paper presents a pipeline for the reconstruction and tracking of the 3D motion of the *ex vivo*, intact, left lobe of a rat lung, as a result of changes in pressure. Model-based 3D reconstruction of the lungs enabled corresponding points to be found between camera views of the fresh lungs. From these, the 3D shape of a patch of the imaged lung could be determined.

The 3D reconstruction of the fresh lung patch in this study was completed with three cameras across 21 pressure steps, encompassing a total pressure change of 317 Pa. The 317 Pa pressure increase resulted in the total mean magnitude of the motion of the lung being 525.7 μm.

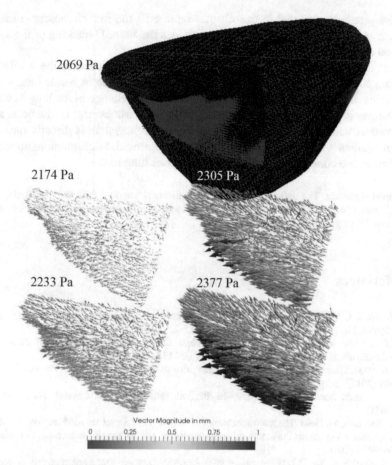

Fig. 10 3D location of the fresh lung surface tracked during inflation. The first frame is shown overlaid on the quadratic Lagrange mesh

This study shows that the 3D reconstruction of the surface of the lungs, using only intrinsic features, is a viable approach to determine 3D shape. A prior 3D mesh was generated from a micro-CT reconstruction of a fixed lung. This mesh was aligned with sparse stereoscopic points identified using a combination of laser line identification and boundary identification on the fresh lung in the stereo-imaging rig. It was shown in this study that a combination of laser line and boundary point identification was sufficient to align the stereoscopic data with the mesh. A model-based reconstruction approach was then used to map texture information across camera views to generate a dense set of corresponding 3D points on the lung surface.

The reconstruction in this study focused on using three cameras to reconstruct a patch of the lung. This demonstrated the feasibility of using such a pipeline for the reconstruction and tracking of fresh lung tissue across a range of pressures without the need for additional surface markers.

The pipeline presented in this chapter represents the first stereoscopic imaging of *ex vivo* lungs. In addition, this work provides the first 3D tracking of the surface motion of the lungs using only intrinsic features.

As part of future work, we aim to extend the reconstruction to the whole lung, making use of all 12 cameras. This will enable 3D tracking of whole lung motion. From this, it will be possible to determine the volume change in the lung as a result of changes in pressure. This will, in turn, enable the assessment of the accuracy of the reconstruction, as volume change in the inflation system is directly measured. Future studies will apply these methods of measuring 3D deformations to identify and model the constitutive properties of the intact lung tissue.

Acknowledgments The authors are grateful for financial support from the Royal Society of New Zealand Marsden Fund, the Medical Technologies Centre of Research Excellence, and the Auckland Bioengineering Institute.

References

1. E. Roan, C.M. Waters, What do we know about mechanical strain in lung alveoli? Am. J. Physiol. Lung Cell. Mol. Physiol. **301**(5), L625–L635 (2011)
2. J.C. Debes, Y.C. Fung, Effect of temperature on the biaxial mechanics of excised lung parenchyma of the dog. J. Appl. Physiol. **73**(3), 1171–1180 (1992)
3. J.B. West, History of respiratory mechanics prior to world war II. Compr. Physiol. **2**(1), 609–619 (2012)
4. W. Mitzner, Mechanics of the lung in the 20th century. Compr. Physiol. **1**(4), 2009–2027 (2011)
5. L. Knudsen, M. Ochs, The micromechanics of lung alveoli: Structure and function of surfactant and tissue components the structural components for gas exchange. Histochem. Cell Biol. **150**, 661–676 (2018)
6. J.S. Kerr, S.Y. Yu, D.J. Riley, Strain specific respiratory air space enlargement in aged rats. Exp. Gerontol. **25**(6), 563–574 (1990)
7. S. Richardson et al., Towards a real-time full-field stereoscopic imaging system for tracking lung surface deformation under pressure controlled ventilation, in *Computational Biomechanics for Medicine*, (Springer International Publishing, Cham, 2019), pp. 119–130
8. A. HajiRassouliha, *A Toolbox for Precise and Robust Deformation Measurement*. The University of Auckland, 2017
9. T. P. Babarenda Gamage, *Constitutive Parameter Identifiability and the Design of Experiments for Applications in Breast Biomechanics*. The University of Auckland, 2016
10. C.C.W. Hsia, D.M. Hyde, M. Ochs, E.R. Weibel, ATS/ERS Joint Task Force on Quantitative Assessment of Lung Structure, An official research policy statement of the American Thoracic Society/European Respiratory Society: Standards for quantitative assessment of lung structure. Am. J. Respir. Crit. Care Med. **181**(4), 394–418 (2010)
11. I. Eltoum, J. Fredenburgh, R.B. Myers, W.E. Grizzle, Introduction to the theory and practice of fixation of tissues. J. Histotechnol. **24**(3), 173–190 (2013)
12. J. Zhang, J. Hislop-Jambrich, T.F. Besier, Predictive statistical models of baseline variations in 3-D femoral cortex morphology. Med. Eng. Phys. **38**(5), 450–457 (2016)
13. A. HajiRassouliha, A.J. Taberner, M.P. Nash, P.M.F. Nielsen, Subpixel phase-based image registration using Savitzky-Golay differentiators in gradient-correlation. Comput. Vis. Image Underst. **170**(170), 28–39 (2018)

A Flux-Conservative Finite Difference Scheme for Anisotropic Bioelectric Problems

George C. Bourantas, Benjamin F. Zwick, Simon K. Warfield,
Damon E. Hyde, Adam Wittek, and Karol Miller

Abstract We present a flux-conservative finite difference (FCFD) scheme for solving inhomogeneous anisotropic bioelectric problems. The method applies directly on the raw medical image data without the need for sophisticated image analysis algorithms to define interfaces between materials with different electrical conductivities. We demonstrate the accuracy of the method by comparison with analytical solution. Results for a patient-specific head model highlight the applicability of the method.

Keywords Flux-conservative finite difference · Anisotropic electrical conductivity · Bioelectric field · Epilepsy · EEG · Patient-specific head model

1 Introduction

Epilepsy is a neurological condition of recurrent or unprovoked seizures that is thought to affect 1% of children [1]. Antiepileptic drugs serve as the primary treatment [2]. Treatment strategy relies on two key issues. First, the quality of life of an epileptic patient fails to improve until the permanent cessation of seizures. Second, one third of patients experience drug resistance [2, 3]. Surgery to remove or alter the region of the brain where seizures originate is recommended to patients who fail to respond to antiepileptic drug therapy [4].

G. C. Bourantas · B. F. Zwick
Intelligent Systems for Medicine Laboratory, Department of Mechanical Engineering, The University of Western Australia, Perth, WA, Australia

S. K. Warfield · D. E. Hyde
Computational Radiology Laboratory, Children's Hospital Boston, and Harvard Medical School, Boston, MA, USA

A. Wittek (✉) · K. Miller
Intelligent Systems for Medicine Laboratory, The University of Western Australia, Perth, WA, Australia
e-mail: adam.wittek@uwa.edu.au

© Springer Nature Switzerland AG 2020
K. Miller et al. (eds.), *Computational Biomechanics for Medicine*,
https://doi.org/10.1007/978-3-030-42428-2_9

Around 100,000–500,000 patients in the United States of America with drug-resistant epilepsy are surgical candidates each year [2]. However, due to the high risk associated with the surgical procedure, less than 1% of patients are treated this way [2]. For surgical epileptic seizure management, there are two realistic options available: focal resection; or disconnection of the epileptogenic cortex [3]. Of these two options, only complete focal resection of the epileptic lesion offers the possibility of eliminating seizures.

Success of the surgical intervention depends on the ability to accurately identify the seizure onset zone (SOZ), which is to be resected. Intracranial electrodes help to identify the SOZ and map eloquent areas of the brain [5]. Currently, the clinical standard for identifying the SOZ are invasive electroencephalography (iEEG) grids and strips, or stereo-EEG (sEEG) electrodes, deployed stereotactically through holes in the skull [6]. The iEEG or sEEG data recorded during the day is collected and manually interpreted by expert neurophysiologists to identify the electrode(s) most implicated in seizure onset.

Patients (usually young) unable to tolerate conscious cortical mapping for resection are candidates for intracranial electrode-mediated extra-operative mapping [3]. The aim of this mapping is to identify the epileptogenic zone. This zone, which is characterized by low-voltage, fast-current neuronal activity, represents the minimum amount of cortex that must be resected to eliminate seizures [7]. Magnetic resonance images (MRIs) are routinely used to determine the distribution of various tissue types throughout the brain. EEGs are used to localize the SOZ and the corresponding area of the brain, which is known as the eloquent cortex [8]. Following the initial MRI, patients undergo a craniotomy to implant intracranial EEG electrodes to the edges of the dura [3]. A low-resolution computed tomography (CT) scan is then used to locate the electrodes within the deformed brain [9].

Source localization of the epileptic zone can be enhanced using computational methods combined with the available imaging modalities. The pre-surgical planning capabilities for resection of the epileptogenic cortex will then be more accurate. Calculating the voltage distribution throughout a patient-specific head model is a key component of the forward problem of EEG source localization. The forward problem has been solved in previous studies using a preoperative brain model [4, 7, 10, 11]. However, a more efficient method for computing the voltage terms is required for patient-specific applications and efficient implementation into the clinical workflow. Previous studies employed finite element methods or boundary element methods to localize the epileptogenic source [12–14]. These methods, however, are limited by their dependence on meshes that sufficiently capture the discontinuity of electrical conductivities between the differing media within the head [15]. Another issue with mesh-based methods is their reliance on pre-determined boundary positions at patient-specific conductivity interfaces within the cortex. Although a high-quality mesh will provide a simple solution to the forward problem, it requires an experienced analyst, thereby decreasing the practicality of implementing this technology into clinical practice.

In this study, we apply the flux-conservative finite difference (FCFD) method to numerically solve the forward problem of EEG source localization. The bioelectric problem is described by a set of partial differential equations. FCFD method discretizes these equations into a system of linear algebraic equations. The numerical solution of the linearized system determines the electric potential distribution throughout a patient-specific conducting volume (head model). The FCFD method applies to the rectangular grid of material properties extracted from patient data. This eliminates image segmentation and meshing that is required in mesh-based methods. The conductivity assigned to each node is used to form a system of linear equations that is then solved to compute the voltage term. We apply an anisotropic tensor for the electrical conductivity. We solve a simple problem with analytical solution to highlight the accuracy of the proposed scheme before applying it to a patient-specific head model of an epilepsy patient.

2 Methods

2.1 Electromagnetic Modeling Using the Flux-Conservative Finite Difference Method

2.1.1 Governing Equations

Source localization methods usually use a linear model, often called leadfield matrix, to correlate measured electrode voltages to their cerebral current sources. Computing the leadfield matrix requires the numerical solution of Maxwell's equations within the head (conducting medium). Since the frequencies employed for EEG are typically less than 100 Hz, transient signals are negligible, and the quasi-static approximation can be employed [4]. Therefore, the relationship between current sources and the induced voltage field is given as:

$$\nabla \cdot \left(\bar{\bar{\sigma}} \left(\mathbf{x} \right) \nabla \Phi \left(\mathbf{x} \right) \right) = \nabla \cdot \mathbf{J} \left(\mathbf{x} \right) \tag{1}$$

with $\Phi(\mathbf{x})$ being the voltage potential at location $\mathbf{x}$ in the spatial domain Ω, $\bar{\bar{\sigma}}(\mathbf{x})$ the spatially varying conductance of the volume, and $\mathbf{J}(\mathbf{x})$ the current source density at the nodes of the volume. The inhomogeneous conductivity tensor $\bar{\bar{\sigma}}(\mathbf{x})$ can be represented by a 3×3 matrix as

$$\bar{\bar{\sigma}} \left(\mathbf{x} \right) = \begin{bmatrix} \sigma_{xx} & \sigma_{xy} & \sigma_{xz} \\ \sigma_{yx} & \sigma_{yy} & \sigma_{yz} \\ \sigma_{zx} & \sigma_{zy} & \sigma_{zz} \end{bmatrix} \tag{2}$$

while the left-hand side of Eq. (1) in its expanded form is given as

$$\nabla \cdot \left(\overline{\overline{\sigma}} \left(\mathbf{x} \right) \nabla \Phi \left(\mathbf{x} \right) \right) = \frac{\partial}{\partial x} \left(\sigma_{xx} \frac{\partial \Phi}{\partial x} + \sigma_{xy} \frac{\partial \Phi}{\partial y} + \sigma_{xz} \frac{\partial \Phi}{\partial z} \right)$$

$$+ \frac{\partial}{\partial y} \left(\sigma_{yx} \frac{\partial \Phi}{\partial x} + \sigma_{yy} \frac{\partial \Phi}{\partial y} + \sigma_{yz} \frac{\partial \Phi}{\partial z} \right) + \frac{\partial}{\partial z} \left(\sigma_{zx} \frac{\partial \Phi}{\partial x} + \sigma_{zy} \frac{\partial \Phi}{\partial y} + \sigma_{zz} \frac{\partial \Phi}{\partial z} \right)$$

$$(3)$$

Using the Taylor series expansion and applying the flux-conservative finite difference scheme we can compute the spatial derivatives of Eq. (3). In the FCFD method, we can efficiently and accurately deal with the anisotropy and the discontinuities in the electrical conductance of the different materials (e.g. bone, soft tissue) in the brain. In the FCFD method we do not apply the chain rule in the computation of the spatial derivatives in Eq. (3), instead we treat the terms in the parenthesis for the spatial derivatives $\frac{\partial}{\partial x}, \frac{\partial}{\partial y}, \frac{\partial}{\partial z}$ as the unknow field functions. Therefore, the typical methodology applied in the classical FD methods is extended to account for the anisotropy of the field variables.

2.2   Flux-Conservative Finite Difference Method

The FD method works efficiently on Cartesian grids (that can be directly obtained from DICOM images) and computes the nonlinear convective term $\nabla \cdot \left(\overline{\overline{\sigma}} \left(\mathbf{x} \right) \nabla \Phi \left(\mathbf{x} \right) \right)$ by applying a flux-conservative scheme. All Flux-Conservative FD formulations give a nodal equation for the potential field $\Phi(\mathbf{x})$ at each node of the grid. The nodal equations finally form a linear algebraic system which can be solved using direct or iterative solvers (for FD method several robust solvers exist).

This scheme computes spatial derivatives for the electric field using the stencil defined in Fig. 1. This is identical to the classical FD stencil except that in the FCFD stencil, fluxes in the fictitious grid points $((i + 1/2, j), (i-1/2, j), (i, j + 1/2), (i, j + 1/2))$ are preserved. Computation of the diffusion term at the grid points $((i, j), (i-1, j), (i, j + 1), (i, j + 1), (i, j-1))$ will lead to an erroneous non-conservative FD formulation. Application of classical (non-conservative) FD stencil by directly applying the chain rule to compute the spatial derivatives of the convective term will lead to incorrect calculation of fluxes.

Using the flux conservative approach, the terms at the central node (i, j, k) of the stencil shown in Fig. 1 can be written (for the x coordinate) as

$$\frac{\partial Q^x}{\partial x} = \frac{Q^x_{\left(1 + \frac{1}{2}, j, k \right)} - Q^x_{\left(1 - \frac{1}{2}, j, k \right)}}{h_x} \tag{4}$$

where

$$Q^x = \sigma_{xx} \frac{\partial \Phi}{\partial x} + \sigma_{xy} \frac{\partial \Phi}{\partial y} + \sigma_{xz} \frac{\partial \Phi}{\partial z} \tag{5}$$

Fig. 1 The 3D stencil configuration used in the flux-conservative finite difference method

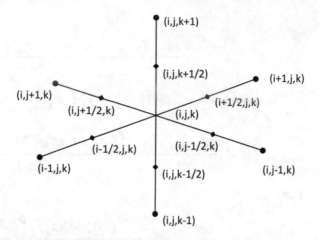

We compute the terms σ_{xx}, $\Phi_{,x}$, σ_{xy}, $\Phi_{,y}$, σ_{xz} and $\Phi_{,z}$ on the off-grid nodes $\left(i+\frac{1}{2}, j, k\right)$ and $\left(i-\frac{1}{2}, j, k\right)$. The electrical conductance σ_{xx}, σ_{xy}, σ_{xz} values are not defined on the off-grid nodes. Instead, they are computed using interpolating/approximating methods such as arithmetic averaging of the known values for the electrical conductance on the grid nodes, or the harmonic average. The former applies for the case of the σ_{xx} electrical conductance (the same applies for σ_{xy} and σ_{xz}) as

$$\sigma_{xx}\left(i+\frac{1}{2}, j, k\right) = \frac{\sigma_{xx\,(i+1, j, k)} + \sigma_{xx\,(i, j, k)}}{2} \tag{6}$$

while the latter is written as

$$\sigma_{xx}\left(i+\frac{1}{2}, j, k\right) = \frac{2\sigma_{xx\,(i+1, j)}\sigma_{xx\,(i, j)}}{\sigma_{xx\,(i+1, j)} + \sigma_{xx\,(i, j)}} \tag{7}$$

The two approaches, despite their success in delivering reliable results, may result in decreased accuracy for the numerical solution when steep gradients in material properties (higher than 6 orders of magnitude) are present. This is because only the two nodes adjacent to the fictitious point are used in the computation, disregarding all the other nodes in the close vicinity. High order methods can be used to provide more accurate results but these increase the computational cost.

Furthermore, we need to compute the spatial derivatives of the electrical potential $\Phi(\mathbf{x})$ on the off-grid nodes. The derivative $\Phi_{,x}$ on the $\left(i+\frac{1}{2}, j, k\right)$ and $\left(i-\frac{1}{2}, j, k\right)$ nodes is given as

$$\frac{\partial \Phi_{(i+1/2, j, k)}}{\partial x} = \frac{\Phi_{(i+1, j, k)} - \Phi_{(i, j, k)}}{h_x} \tag{8}$$

and

$$\frac{\partial \Phi_{(i-1/2,j,k)}}{\partial x} = \frac{\Phi_{(i,j,k)} - \Phi_{(i-1,j,k)}}{h_x} \tag{9}$$

The derivative $\Phi_{,y}$ on the $\left(i + \frac{1}{2}, j, k\right)$ and $\left(i - \frac{1}{2}, j, k\right)$ nodes is given as

$$\frac{\partial \Phi_{(i+1/2,j,k)}}{\partial y} = \frac{\Phi_{\left(i+\frac{1}{2},j+1/2,k\right)} - \Phi_{\left(i+\frac{1}{2},j-1/2,k\right)}}{h_y} \tag{10}$$

and

$$\frac{\partial \Phi_{(i-1/2,j,k)}}{\partial y} = \frac{\Phi_{\left(i-\frac{1}{2},j+1/2,k\right)} - \Phi_{\left(i-\frac{1}{2},j-1/2,k\right)}}{h_y} \tag{11}$$

where

$$\Phi_{\left(i+\frac{1}{2},j+1/2,k\right)} = \frac{\Phi_{(i,j,k)} + \Phi_{(i+1,j,k)} + \Phi_{(i+1,j+1,k)} + \Phi_{(i,j+1,k)}}{4} \tag{12}$$

$$\Phi_{\left(i+\frac{1}{2},j-1/2,k\right)} = \frac{\Phi_{(i,j,k)} + \Phi_{(i+1,j,k)} + \Phi_{(i+1,j-1,k)} + \Phi_{(i,j-1,k)}}{4} \tag{13}$$

$$\Phi_{\left(i-\frac{1}{2},j+1/2,k\right)} = \frac{\Phi_{(i,j,k)} + \Phi_{(i,j+1,k)} + \Phi_{(i-1,j+1,k)} + \Phi_{(i-1,j,k)}}{4} \tag{14}$$

$$\Phi_{\left(i-\frac{1}{2},j-1/2,k\right)} = \frac{\Phi_{(i,j,k)} + \Phi_{(i,j-1,k)} + \Phi_{(i-1,j,k)} + \Phi_{(i-1,j-1,k)}}{4} \tag{15}$$

Finally, the derivative $\Phi_{,z}$ on the $\left(i + \frac{1}{2}, j, k\right)$ and $\left(i - \frac{1}{2}, j, k\right)$ nodes is given as

$$\frac{\partial \Phi_{(i+1/2,j,k)}}{\partial z} = \frac{\Phi_{\left(i+\frac{1}{2},j,k+1/2\right)} - \Phi_{\left(i+\frac{1}{2},j,k-1/2\right)}}{h_z} \tag{16}$$

and

$$\frac{\partial \Phi_{(i-1/2,j,k)}}{\partial z} = \frac{\Phi_{\left(i-\frac{1}{2},j,k+1/2\right)} - \Phi_{\left(i-\frac{1}{2},j,k-1/2\right)}}{h_z} \tag{17}$$

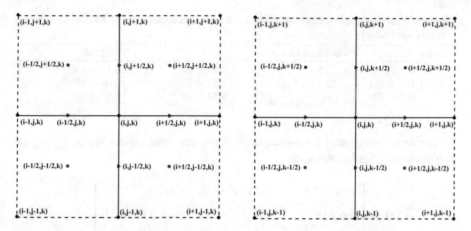

Fig. 2 The 3D stencil configuration used in the flux-conservative finite difference method

where

$$\Phi_{\left(i+\frac{1}{2},j+1/2,k\right)} = \frac{\Phi_{(i,j,k)} + \Phi_{(i+1,j,k)} + \Phi_{(i+1,j+1,k)} + \Phi_{(i,j+1,k)}}{4} \tag{18}$$

Consequently, for computing the partial derivative with respect to x for the Q^x term, eight neighbors are involved. Figure 2 shows the grid nodes used in the computation of the term $\frac{\partial}{\partial x}\left(\sigma_{xx}\frac{\partial\Phi}{\partial x} + \sigma_{xy}\frac{\partial\Phi}{\partial y} + \sigma_{xz}\frac{\partial\Phi}{\partial z}\right)$.

The same procedure applies for the other two partial derivatives $\frac{\partial}{\partial y}\left(\sigma_{yx}\frac{\partial\Phi}{\partial x} + \sigma_{yy}\frac{\partial\Phi}{\partial y} + \sigma_{yz}\frac{\partial\Phi}{\partial z}\right)$ and $\frac{\partial}{\partial z}\left(\sigma_{zx}\frac{\partial\Phi}{\partial x} + \sigma_{zy}\frac{\partial\Phi}{\partial y} + \sigma_{zz}\frac{\partial\Phi}{\partial z}\right)$ in Eq. (3). Therefore, 27 neighboring nodes form the stencil for computing the left-hand side in Eq. (1). The right-hand side ($\nabla \cdot \boldsymbol{J}(\boldsymbol{x})$) is also defined on the grid nodes and can be defined as a continuous function, discretized over the nodes, or as point sources.

3   Results

3.1   Verification of the FCFD Scheme

To demonstrate the accuracy of the proposed FCFD scheme we solve the Laplace equation for an inhomogeneous anisotropic medium in a unit volume box. The problem has an analytical solution of the form

$$\Phi(\boldsymbol{x}) = e^{x+y+z} \tag{19}$$

Table 1 Maximum relative error and normalized root mean square error (NRMSE) for increasing grid resolution

Grid resolution	Solution time (s)	L_∞	NRMSE
$51 \times 51 \times 51$	13	2.21×10^{-5}	2.13×10^{-6}
$101 \times 101 \times 101$	228	1.02×10^{-5}	7.08×10^{-7}
$201 \times 201 \times 201$	3363	6.72×10^{-6}	6.59×10^{-7}

For an inhomogeneous anisotropic medium, the conductivity tensor giving the analytical solution has the form

$$\overline{\overline{\sigma}}\,(\mathbf{x}) = \begin{bmatrix} e^{x+y+z} & -0.25e^{x+y+z} & -0.75e^{x+y+z} \\ -0.25e^{x+y+z} & 1.5e^{x+y+z} & -1.25e^{x+y+z} \\ -0.75e^{x+y+z} & -1.25e^{x+y+z} & 2e^{x+y+z} \end{bmatrix} \qquad (20)$$

We apply Dirichlet boundary conditions on the boundary nodes, according to the analytical solution (Eq. 19).

The linear system of the Laplace equation can be solved using direct or iterative solvers. The former are extremely accurate but have memory limitations, especially for 3D problems with large number of nodes. The latter do not always converge but are extremely efficient and have less computational cost compared to direct solvers. For the systems used in the present study, we use the minimum residual method, which applies to nonsymmetric systems. We used an Intel i7 quad core processor with 16 GB RAM for our simulations.

We compare the numerical solution against the analytical one using the Normalized Root Mean Square Error defined as $NRMSE = \dfrac{\sqrt{\frac{1}{N} \sum_{i=1}^{N} \left(u_i^{numerical} - u_i^{analytical} \right)^2}}{u_{max}^{analytical} - u_{min}^{analytical}}$.
To study the convergence of the solution, we used successively denser grids starting from $51 \times 51 \times 51$ up to $201 \times 201 \times 201$.

The results (Table 1) suggest that both the maximum relative error and NRMSE will converge to zero as the number of nodes increases, confirming the accuracy of the FCFD scheme for solving anisotropic, three-dimensional, bioelectric field problems. Figure 3 shows the potential distribution computed by the analytical solution at plane $z = 0.5$ and a histogram displaying the differences, node by node, of the numerical solution with the analytical one for a grid resolution of 101^3.

For source localization, the computational time needed to solve the forward problem is crucial because multiple forward problems must be solved. Therefore, the accuracy and efficiency provided from the proposed scheme makes it a strong candidate to be used in clinical practice.

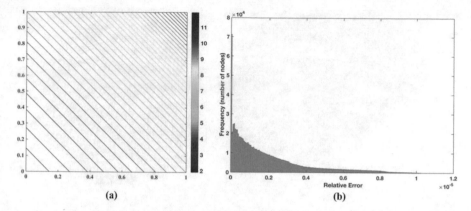

Fig. 3 Axial view of the (**a**) numerical solution and (**b**) histogram of the differences with the analytical solution using the flux-conservative finite difference method for the for inhomogeneous anisotropic medium verification problem

3.2 Patient-Specific Head Model

In this section, we apply the FCFD method to a patient-specific head model of a five-year old epilepsy patient. The electrical conductivities were extracted from the patient's diffusion-weighted MRI using the method described in [16]. A node was assigned to the corner of each voxel to create a $160 \times 192 \times 192$ grid comprised of 5,898,240 points. An anisotropic conductivity was assigned to all nodes inside the conducting volume. The three-dimensional finite difference brain volume was comprised of white and grey matter, as well as cerebrospinal fluid and air. We model air, grey matter and cerebrospinal fluid conductivities using isotropic tensors, while white matter fibers were assigned anisotropic tensors. Using the Cartesian grid (voxels) directly from the raw data we avoid the need for image segmentation to assign constitutive properties (Fig. 4).

We compute the electric potential distribution throughout the brain volume by applying the point electrode model. We selected electrodes as the source and sink. We apply a current of 1 A at the source, and we remove 1 A at the sink. In the presence of any external current source, Poisson's equation (Eq. 10) governs the potential distribution within the head volume incorporating anisotropic conductivity. At the boundaries, we enforce Neumann boundary conditions (Eq. 2). We numerically solve the linear system of equations using the minimum residual method. We model air using an isotropic conductivity of 10^{-9} (S/m), assigning this to all voxels outside of the head volume. This is demonstrated in Fig. 5a–c as the voltage approaches zero outside the skull-air interface boundary.

Figure 5 shows the electric potential distribution throughout the brain in the axial, coronal and sagittal planes. These slices center around the midpoint of the preselected source/sink configuration to best illustrate the voltage distribution (we positioned the source at (143, 114, 101) and the sink at (110, 102, 104)). As

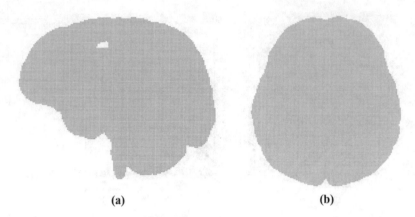

(a)   (b)

Fig. 4 (a) Sagittal and (b) axial view of the brain raw data

expected, the source and sink generate a voltage inside the conducting volume that is greatest close to the corresponding electrodes and approaches zero as the distance from these regions increases.

4 Conclusion

In this study, we successfully applied the FCFD method to numerically solve the bioelectric problem to obtain the voltage distribution throughout the head. We first applied the FCFD method to a simple problem with an analytic solution. Following verification, the proposed scheme has been applied to a patient-specific head model (created using raw medical image data) to compute the electric potential distribution throughout the conducting volume for a specified source/sink configuration.

The accuracy of the patient-specific head model may be improved by using a complete electrode model instead of the point-electrode model used in the present study. The complete electrode model incorporates the size of the electrodes, their shape and the contact impedance, providing a better approximation of the electrode-tissue interface. With the point-electrode model, currents in the electrodes are not considered in the numerical solution. Therefore, the voltages close to the electrodes are of greater amplitude compared to those expected in real-world cases.

Successful application of the proposed scheme enhances current pre-surgical planning capabilities for resection of the epileptogenic cortex. In contrast to traditional mesh-based methods such as the finite element and boundary element methods, with our method there is no need for image segmentation and mesh generation.

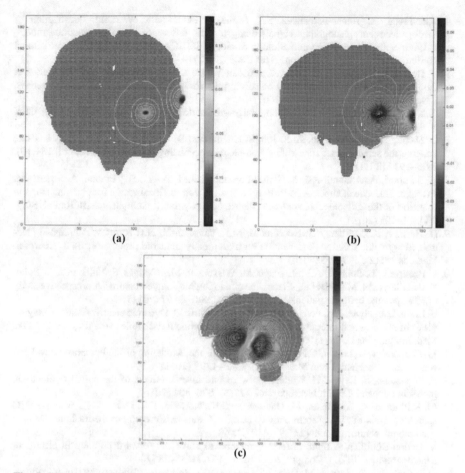

Fig. 5 Electric potential distribution throughout the brain in (**a**) axial plane 103, (**b**) coronal plane 108, and (**c**) sagittal plane 127

Acknowledgements This research was supported by the Australian Government through the Australian Research Council's *Discovery Projects* funding scheme (project DP160100714) and National Health and Medical Research Council Project Grant APP1162030.

References

1. O. Devinsky, *Epilepsy: A Patient and Family Guide* (Demos Medical Publishing, New York, 2007)
2. V.M. Ravindra, M.T. Sweney, R.J. Bollo, Recent developments in the surgical management of paediatric epilepsy. Arch. Dis. Child. **102**(8), 760–776 (2017)
3. A.M. Husain, *Practical Epilepsy*, 1st edn. (Springer Publishing Company, New York, 2015)

4. D.E. Hyde, X. Tomas-Fernandez, S.S. Stone, J. Peters, S.K. Warfield, Localization of stereo- electroencephalography signals using a finite difference complete electrode model, in Engineering in Medicine and Biology Society (EMBC), 2017 39th Annual International Conference of the IEEE, IEEE, pp. 3600–3603 (2017)
5. H. Hallez, B. Vanrumste, R. Grech, J. Muscat, W. De Clercq, A. Vergult, Y. D'Asseler, K.P. Camilleri, S.G. Fabri, et al., Review on solving the forward problem in EEG source analysis. J. Neuroeng. Rehabil. 4(1), 46–75 (2007)
6. R.A. Vega, K.L. Holloway, P.S. Larson, Image-guided deep brain stimulation. Neurosurg. Clin. N. Am. 25, 159–172 (2014)
7. O. David, T. Blauwblomme, A.-S. Job, S. Chabardès, D. Hoffmann, L. Minotti, P. Kahane, Imaging the seizure onset zone with stereo-electroencephalography. Brain J. Neurol. 134(10), 2898–2911 (2011)
8. V. Taimouri, A. Akhondi-Asl, X. Tomas-Fernandez, J.M. Peters, S.P. Prabhu, A. Poduri, M. Takeoka, T. Loddenkemper, A.M.R. Bergin, et al., Electrode localization for planning surgical resection of the epileptogenic zone in pediatric epilepsy. Int. J. Comput. Assist. Radiol. Surg. 9(1), 91–105 (2014)
9. D. Hermes, K.J. Miller, H.J. Noordmans, M.J. Vansteensel, N.F. Ramsey, Automated electrocorticographic electrode localization on individually rendered brain surfaces. J. Neurosci. Methods 185(2), 293–298 (2010)
10. V. Brodbeck, L. Spinelli, A.M. Lascano, M. Wissmeier, M.-I. Vargas, S. Vulliemoz, C. Pollo, K. Schaller, C.M. Michel, et al., Electroencephalographic source imaging: A prospective study of 152 operated epileptic patients. Brain 134(10), 2887–2897 (2011)
11. B. Erem, D.E. Hyde, J.M. Peters, F.H. Duffy, S.K. Warfield, Dynamic electrical source imaging (desi) of seizures and interictal epileptic discharges without ensemble averaging. IEEE Trans. Med. Imaging 36(1), 98–110 (2017)
12. M. Dannhauer, B. Lanfer, C.H. Wolters, T.R. Knösche, Modeling of the human skull in EEG source analysis. Hum. Brain Mapp. 32(9), 1383–1399 (2011)
13. S. Pursiainen, S. Lew, C.H. Wolters, Forward and inverse effects of the complete electrode model in neonatal EEG. J. Neurophysiol. 117(3), 876–884 (2016)
14. M. Rullmann, A. Anwander, M. Dannhauer, S.K. Warfield, F.H. Duffy, C.H. Wolters, EEG source analysis of epileptiform activity using a 1 mm anisotropic hexahedra finite element head model. NeuroImage 44(2), 399–410 (2009)
15. V. Hédou-Rouillier, A finite difference method to solve the forward problem in electroencephalography (EEG). J. Comput. Appl. Math. 167(1), 35–58 (2004)
16. D.S. Tuch, V.J. Wedeen, A.M. Dale, J.S. George, J.W. Belliveau, Conductivity tensor mapping of the human brain using diffusion tensor MRI. Proc. Natl. Acad. Sci. 98(20), 11697–11701 (2001)

A Fast Method of Virtual Stent Graft Deployment for Computer Assisted EVAR

Aymeric Pionteck, Baptiste Pierrat, Sébastien Gorges, Jean-Noël Albertini, and Stéphane Avril

Abstract In this paper we introduce a new method simulating stent graft deployment for assisting endovascular repair of abdominal aortic aneurysms. The method relies on intraoperative images coupled with mechanical models. A multi-step algorithm has been developed to increase the reliability of simulations. The first step predicts the position of the stent graft within the aorta. The second step is an axisymmetric geometric reconstruction of each individual stent. The third step minimizes the rotation of each stent around its main axis. Finally, the last step virtually deploys each stent within a deployment box extracted from the preoperative CT scan. A proof of concept is performed on a patient. The accuracy is compatible with the clinical threshold of 3 mm: the average distance between target and simulated stents is 1.73 ± 0.37 mm. Fenestrations of the stent-graft are reconstructed with a maximum error of less than 2.5 mm, which enables a secure catheterization of secondary arteries. In summary, the method is able to assist EVAR practitioners by providing all necessary information for a fast and accurate stent graft positioning, combining intraoperative data and a mechanical model in a very low cost framework.

A. Pionteck
Mines Saint-Etienne, University of Lyon, Jean Monnet University, INSERM, U1059 Sainbiose, Centre CIS, Saint-Etienne, France

THALES, Microwave & Imaging Sub-Systems, Moirans, France

B. Pierrat · S. Avril (✉)
Mines Saint-Etienne, University of Lyon, Jean Monnet University, INSERM, U1059 Sainbiose, Centre CIS, Saint-Etienne, France
e-mail: avril@emse.fr

S. Gorges
THALES, Microwave & Imaging Sub-Systems, Moirans, France

J.-N. Albertini
Jean Monnet University, INSERM, U1059 Sainbiose and University Hospital of Saint-Etienne, Saint-Etienne, France

© Springer Nature Switzerland AG 2020
K. Miller et al. (eds.), *Computational Biomechanics for Medicine*,
https://doi.org/10.1007/978-3-030-42428-2_10

Keywords Abdominal aortic aneurysms (AAA) · Endovascular aneurysm repair (EVAR) · Finite element analysis (FEA) · Augmented reality · Real-time simulation · Computed tomography · Stent-graft · Computer assisted surgery

1 Introduction

Abdominal aortic aneurysm (AAA) is a frequent asymptomatic pathology that results in abnormal local deformation of the aorta. Each year, aneurysm ruptures are responsible for 10,000 deaths in the United States [1]. Clinical monitoring of the evolution of the aneurysmal sac diameter is used to decide whether an intervention is necessary [1, 2]. Two options are available: conventional open surgery or endovascular surgery (EVAR). Endovascular surgery is associated with a lower mortality rate (1.5%) than open surgery (4.6%), although long-term mortality is similar [3, 4].

During EVAR, the surgeon first makes a small incision in the groin to reach the femoral artery. From this incision, tools are introduced to position the stent graft (SG) launcher within the aneurysm. Then the SG is progressively deployed. The success of the intervention depends on the precise positioning of the SG in the artery. In some cases, a fenestrated SG is required if the aneurysm extends beyond the ostia of the renal arteries. In this case, the fenestrations of the SG must be positioned precisely in front of the renal ostium whose diameter is about 5–7 mm. This phase is delicate but essential to allow the catheterization of the secondary arteries and avoid occlusions and post-operative complications [5–8]. The lack of 3D information obliges the surgeon to perform a mental reconstruction of the scene using several images with different incidence angles, which considerably increases the duration of the procedure, as well as the time of exposure to X-rays and the volume of injected contrast products. Therefore, the virtual 3D representation of the tool location and particularly the SG in the aorta is a valuable aid to the surgeon. This would reduce the surgery time and the number of X-ray images required. Moreover, it would reduce the number of postoperative complications, most often related to inaccurate SG positioning. Latest generation systems enable the acquisition of 3D images of the tool during the intervention [9, 10]. Recently, efforts have been made to use biplanar fluoroscopic acquisitions to reconstruct the 3D shape of the device [11–14]. However, all these methods are based on expensive equipment which are not commonplace in all hospitals, usually equipped with simple mobile C-arms. Another solution should therefore be available to obtain a three-dimensional representation of the inserted SG at low cost. Modelling and numerical simulation of SG deployment then appears as essential.

First studies on numerical simulation of SG deployment were based on finite element analyses to study the mechanics of stents and to simulate their deployment in arteries [15–17], integrating different types of constitutive behavior for the different materials of SGs [18–20].

Perrin et al. [21–24] developed a preoperative planning tool to predict the postoperative position of the SG from patient-specific models. Although essential for preoperative planning, these studies have two important limitations with regard to their use as real time assistance for the practitioner: (i) inappropriately long computation time, and (ii) lack of update from intraoperative images. Some studies have focused on reducing the computation time and developed algorithms to simulate stent deployment in "real time". They often rely on simplifications such as modeling vessels as generic tubes, on which the stent armatures are then mapped [25–27]. The Fast Virtual Stenting (FVS) technique was proposed by Larrabide et al. [28]. This technique is based on constrained deformable simple models and can virtually model stent deployment in vessel and aneurysm models. The FVS technique was tested and compared with experimental results [29] and with finite element models [30, 31]. Alternative methods have been proposed, based on mass-spring models [32, 33] or on active contours [34]. Although efficient and fast, most of these models are based on simplified mechanics and can be challenged by complex vascular geometries. In addition, applications focus on preoperative planning, as none of the work mentioned above considered intraoperative images.

A small number of studies integrated information from intraoperative images. For example, Demirci et al. [35] proposed an algorithm to automatically match a 3D model of the SG with an intraoperative 2D image of its structure. Zhou et al. [36], and Zheng et al. [37] introduced a real-time framework to generate the 3D shape of a fenestrated SG from a single 2D fluoroscopic image and position of added radio-opaque markers. These methods have reduced computation times and can accurately represent the deployment of SGs in simple geometries. However, more complex cases cannot be addressed without the use of a mechanical model.

To our best knowledge, no studies have ever combined these different aspects into a single method. Achieving this combination is the objective of the present work, in order to propose a method that can assist EVAR practitioners by providing all necessary information for a fast and accurate SG positioning.

The details of the method are given in this book chapter, first introducing the global algorithm, then describing each step and finally showing a proof of concept for a patient case.

2 Methods

The global algorithm of the method is summarized in Fig. 1. The input data are the 2D intraoperative images from a mobile C-arm and the 3D geometry of the aorta obtained from a preoperative CT scan. The algorithm is divided into four main steps. The first two steps can be combined into a single stage called Stage 1. This stage is essential for the following steps but may reach insufficient accuracy, hence the possible following Stage 2. During the first step of Stage1, barycenters of each stent are positioned in 3D using a FEM model of the SG in the aorta. Then, the stents are geometrically reconstructed during the second step of Stage 1. If necessary, two refining steps are achieved during Stage 2, which is an updating or refining stage. These suplemental steps require a slightly longer calculation time but reach higher accuracy. The first step of Stage 2 consists in recovering the rotation of the stent around its main axis through a minimization loop. The second step of Stage 2 consists in deploying each stent individually.

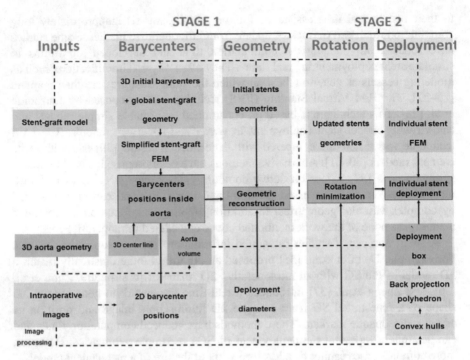

Fig. 1 Schematic of the general algorithm

2.1 Data Acquisition

In this section, we list the input data and describe how they are processed to extract relevant information and feed the simulations. Data available at the beginning of SG reconstruction include the SG model, updated 3D aorta geometry and intraoperative imaging. The SG models are obtained using the method described in [22, 38]. Briefly, stent geometries are obtained from manufacturer specifications and are discretized into finite elements using a dedicated Matlab® routine. All simulations are constrained and guided by intraoperative imaging. To isolate the contour of the stents, a combination of Frangi filters [39] and masks is applied to the image [35]. The Frangi filter is generally used to detect vessels or tubular structures in volumetric image data. The Frangi filter is available in open-source libraries and software such as ITK or ImageJ. The filter includes a measuring scale that allows the isolation of tubular structures of different sizes. By modifying the scale of the filter and combining it with masks, it is possible to extract binary tubular structures of the stents (Fig. 2). Then, the convex hull of each stent is extracted. The two-dimensional coordinates of stent barycenters are simply obtained from the convex hull (Fig. 2). Apparent deployment diameters are also measured. For each stent, the proximal diameter d_p and distal diameter d_d are recorded. They will be used for the further geometric reconstruction of stents.

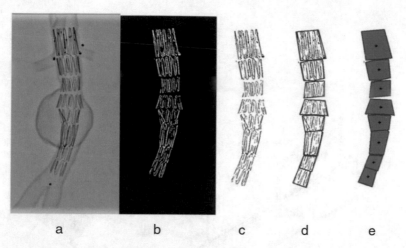

Fig. 2 Example of stent deployment in a 3D printed AAA replica: original image (**a**), stent detection (**b**) and extraction (**c**), convex hull (**d**) and 2D barycenters (**e**)

In the following steps, the SG is virtually positioned in the 3D geometry of the aorta. The aortic geometry, including the centerline and the volume, is previously extracted from the preoperative scan. The volume of the aorta is segmented from the DICOM file of the preoperative scan with a front collision method implemented in VMTK [40]. Then, the centerline is extracted using the Voronoi diagram method, also implemented in VMTK. The geometry of the aorta obtained from the preoperative scans may be slightly different of the aortic geometry at the day of the intervention. Indeed, it can be deformed, especially when stiff guidewires are inserted. The aortic geometry must be updated before simulating SG deployment. To do so, the geometry is rigidly and then non-rigidly registered on the intraoperative images. Several methods are available for this step [41, 42] ([43] under review). In addition, we assume that we know the projection matrix of the C-arm.

2.2 Corotational Euler-Bernoulli Beam Elements

In this section we describe the corotational Euler-Bernoulli beam elements that are used in the following steps to discretize the simplified geometry of the SG in the global positioning step, and then the stents in the individual deployment step. Simulations are carried out with Project Chrono [44, 45]. The details of the theory and implementation of beam elements are described in [46]. We review here the main concepts. Among the different methods that allow simulating large deformations by finite elements, the corotational approach is one of the most versatile as it is based on classical linear finite elements. The corotational approach allows large displacements, but requires that the strains remain small (Fig. 3).

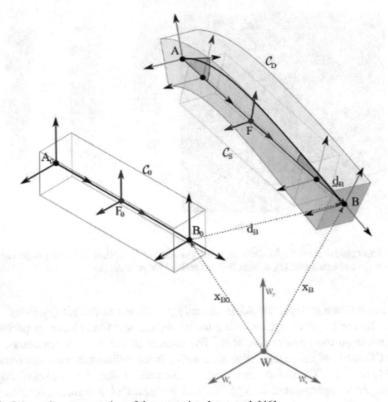

Fig. 3 Schematic representation of the corotational approach [46]

A floating coordinate system **F** follows the deformed element, so that the overall movement in the deformed $\mathbf{C_D}$ state can be assumed to be composed of a large rigid body movement from the reference configuration $\mathbf{C_0}$ to the so-called floating or phantom configuration $\mathbf{C_S}$, times a small local deformation from $\mathbf{C_S}$ to $\mathbf{C_D}$. The underlined symbols represent variables expressed in the floating reference basis **F**. A global tangent stiffness K_e and a global force vector $\mathbf{f_e}$ are derived for each element e, given its local matrix $\underline{K}$, its local force $\underline{f}$ and the rigid body motion of **F** in $\mathbf{C_0}$ to **F** in $\mathbf{C_S}$. At each time step, the position and rotation of **F** are updated.

2.3 Stage 1 (Preliminary Stage)

Stage 1 combines the first two steps of the algorithm: positioning of the stent barycenters in the aorta and axisymmetric reconstruction of the stents. This step can be run in real-time as it has a marginal computational cost. However, it is based on assumptions that may not be fully satisfied in practice. Thus, the output of this preliminary stage will serve as the starting-point for the updating stage presented in

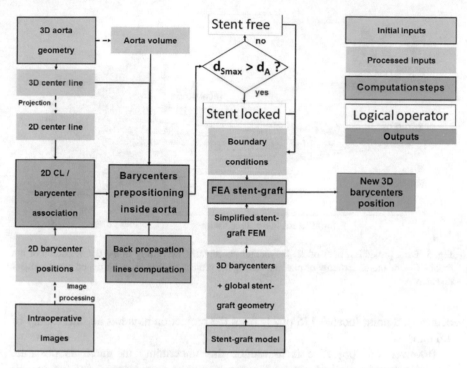

Fig. 4 Overview of the algorithm for 3D positioning of barycenters

the next section. The first step is to recover the global position of the SG inside the aorta. The SG is simulated with a simplified finite element model. The algorithm for positioning barycenters in 3D is summarized in Fig. 4.

2.3.1 Barycenter Positioning

It is very challenging to find the position of the SG directly in the global reference frame from a single image. Indeed, the intrinsic nature of the C-arm conical projection and the discretization in pixels of the detectors lead to a significant incertitude along the projection axis. A pixel from the flat panel can be assimilated to a surface, therefore its back projection geometry is not a line but a pyramidal volume. It is from this volume that the uncertainty on point positioning comes from (Fig. 5). This uncertainty represents the distance along which an object can be moved along the projection axis without moving in the projection plane. This uncertainty depends on the position of the object, i.e. the source-object distance but also the distance from the projection axis. When the pixel gets closer to the projection axis, the uncertainty may tend towards infinity. Close to the edges of the image, the uncertainty becomes lower. For a standard case (source-object distance = 800 mm, source-detector distance = 1300 mm, 750 pixel × 750 pixel

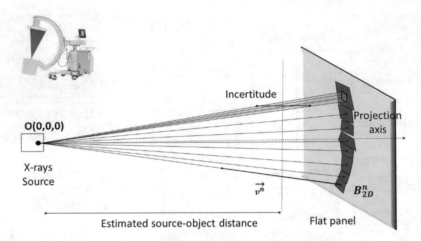

Fig. 5 Back projection lines of 2D barycenter coordinates, according to the configuration of the mobile C-arm and uncertainty of positions along the projection axis at an estimated source-object distance

detector), a point located 175 pixels from the projection axis has an uncertainty of 4.6 mm.

However, our objective is to reduce this uncertainty as much as possible. Accordingly, our method uses the aortic geometry as a support for the overall SG positioning. The first step is to relate each stent barycenter with the closest point of the artery centerline. This association is achieved in two dimensions. The centerline of the aorta is projected according to the same projection parameters as the intraoperative image. Each two-dimensional barycenter is simply associated with the nearest projected centerline point. This associativity is converted into three dimensions, assuming that the nearest 2D point is almost equivalent to the nearest 3D point. Each barycenter is then related to the corresponding centerline position CL (x_{CL}, y_{CL}, z_{CL}).

The next step is to obtain the back-projection lines of each 2D barycenter. For each stent, we know (from previous image processing) the convex hull and the 2D position of its barycenter B_{Im} (U_{Im}, V_{Im}) in the screen frame. The position of the target image in the 3D space, representing the configuration of the X-ray source and the flat panel detector, is known. Thus, each barycenter on the image can be associated with a three-dimensional position B_{2D} (x_{2D}, y_{2D}, z_{2D}). Since we know the projection matrix, we can obtain the corresponding back projection line for each barycenter. These lines are combined with the geometry of the aorta to obtain the global position of the SG in the 3D frame. Equations of the projection lines result from the elementary Cartesian geometry. The X-ray source is at the origin of the global coordinate system. Therefore, all projection lines have point O $(0,0,0)$ in common. The projection parameters are known. Thus, each B_{2D} target point of the image is associated with a back-projection line passing through this point and through the origin (Fig. 5). The normalized vector of line $\overrightarrow{v}$ (v_x, v_y, v_z) is:

$$\vec{\mathbf{v}} = \frac{\overrightarrow{\mathbf{B}_{2D}\mathbf{O}}}{\left\|\overrightarrow{\mathbf{B}_{2D}\mathbf{O}}\right\|} \tag{1}$$

The SG must now be pre-positioned inside the aorta. The coordinates of each three-dimensional barycenter $\mathbf{B}_{3D}$ (x_{3D}, y_{3D}, z_{3D}) have then a back-projection line and have previously been associated with the closest centerline point $\mathbf{CL}$ (x_{CL}, y_{CL}, z_{CL}). We know that each barycenter 3D coordinate is located on its back projection line, but the coordinate p (Eq. 2) of $\mathbf{B}_{3D}$ along its line is **initially** unknown.

$$\mathbf{B}_{3D}\,(x_{3D}, y_{3D}, z_{3D}) = \mathbf{B}_{2D}\,(x_{2D}, y_{2D}, z_{2D}) + p * \vec{\mathbf{v}}\,(v_x, v_y, v_z) \tag{2}$$

As a first approximation, we assign to each barycenter the coordinate p from the nearest centerline point such as $z_{3D} = z_{CL}$. From Eq. 2, we obtain the following system of equations:

$$\begin{cases} x_{3D} = x_{2D} + p * v_x \\ y_{3D} = y_{2D} + p * v_y \\ z_{3D} = z_{2D} + p * v_z \end{cases} \tag{3}$$

Hence:

$$p = \frac{z_{3D} - z_{2D}}{v_z} \tag{4}$$

(x_{3D}, y_{3D}) are calculated by solving the system of Eq. (3). Then, we have a first approximation of the position of barycenters, based on the centerline. In some cases, this approach is insufficient and must be completed using a finite element model of the SG. Figure 6 shows what can happen in the case of a large aneurysm sac. In this case, the centerline follows the shape of the artery. With the approach described above, stents would be positioned in the sac, which is unlikely and mechanically unrealistic.

Therefore, stents that are likely to be badly positioned must be separated from the other ones. In order to define the most precise boundary conditions for the finite element simulation, stents are divided into two categories: free and locked stents. The maximum diameter d_{Smax} of the stent, i.e. the diameter of the fully deployed stent, is compared to the local diameter of the aorta d_A at the associated point of the centerline. If $d_{Smax} > d_A$, the stent is locked. In this case, we assume that the stent is in equilibrium in the artery and that its barycenter is therefore very close to the local center of the artery, and therefore to $\mathbf{CL}$. If $d_{Smax} < d_A$, for example if the stent is in the aneurysm sac, the stent is free. The position of the locked stents is assigned according to the centerline (see previous section). The position of the barycenters of free stents will be calculated using a simplified finite element model of the SG. From the initial three-dimensional geometry of the stent, the 3D position of each stent barycentre is extracted. Barycenters are connected between each other

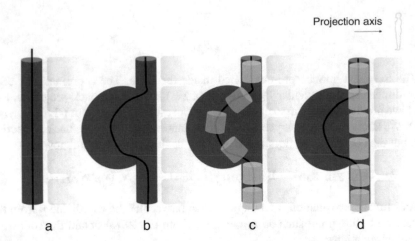

Fig. 6 Error in barycenter positioning due to a deformed centerline: normal centerline (**a**), centerline deformed in an aneurysm sac (**b**), resulting unrealistic stent positioning (blue) in the sac area and realistic positioning (green) in non-deformed sections (**c**), expected actual positioning (**d**)

by corotational Euler-Bernoulli beam elements according to the initial configuration of the SG [46] (Fig. 7). The model is set up using the Project Chrono libraries.

The SG model in its initial configuration is pre-positioned in the aorta. Then, displacements are prescribed onto the locked stents and the resulting displacements of the free stents are calculated. Free stents cannot go outside the aortic lumen. A SG is a tubular structure with a high degree of mechanical inhomogeneity due to the combination of metal stents and textile graft. The SG model is very simplified, reducing the SG model to a succession of beam elements with the same mechanical behavior. The mechanical characteristics of these beams therefore have no physical reality, and have been optimized to ensure the robustness and stability of the model. As the model is subject to successive boundary conditions (back-projection lines, aortic volume), we assume that this simplified model is sufficient for our application, while allowing a very short computation time. The 3D positions of barycenters are finally determined, hence stent orientation.

2.3.2 Geometric Stent Reconstruction

From the updated 3D position of the barycenters, a geometric reconstruction of the stent is performed. The initial geometry of each stent, i.e. the metal structure, is first discretized into a set of points (Fig. 8). Each point is defined as a vector $\vec{V}$ (V_x, V_y, V_z), which originates from the stent barycenter B_{3D} and is expressed in the stent local coordinate system. Initially, the local reference coordinate system is the translated global coordinate system. Therefore, the coordinates $S(x,y,z)$ of the n points of a stent are defined by:

Fig. 7 From the simplified geometry of the SG (**a**), 3D coordinates of barycenters are first extracted (**b**), then barycenters are connected using beam elements (**c**)

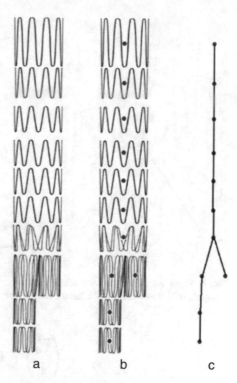

a   b   c

$$S^{i:1 \to n} = B^i_{3D} + \overrightarrow{V^i} \qquad (5)$$

The proximal deployment diameters d_p and distal d_d are measured during the image processing step. Here we assume that the stent deployment is axisymmetric. Thus, the diameter measured in the plane of the image is assumed to be the same in all directions. The local deployment diameter d of the stent is therefore interpolated along its main axis z', initially coinciding with the axis z of the global reference frame (Fig. 8). New reconstruction vectors $\overrightarrow{V_d}\left(V_{dx}, V_{dy}, V_{dz}\right)$ are updated according to the diameter reduction rd such as:

$$rd = 1 - \frac{dSmax - d}{dSmax} \qquad (6)$$

$$\begin{cases} V_{dx} = V_x * rd \\ V_{dy} = V_y * rd \\ V_{dz} = V_z \end{cases} \qquad (7)$$

Reconstruction vectors $\overrightarrow{V_d}$ in the global reference frame is expressed in the global frame with the rotation matrix R according to:

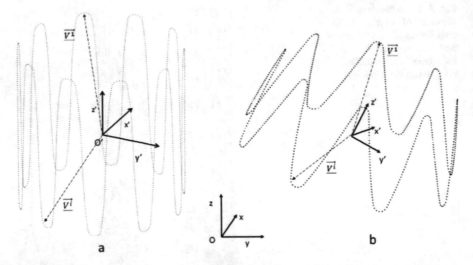

Fig. 8 Stent geometric reconstruction, with the local frame O'(x',y',z') and the global frame O(x,y,z), initial model (**a**) and after reconstruction (**b**)

$$\begin{pmatrix} \vec{V_d} \\ 1 \end{pmatrix} = R \begin{pmatrix} \vec{V_d} \\ 1 \end{pmatrix} \tag{8}$$

Finally, the new position of the stent is calculated with Eq. 5 from the updated vectors $\vec{V_d}$. The SG is eventually reconstructed (Fig. 8).

Positioning and reconstruction of stents is based on two assumptions: the center of gravity of the stents is close to the centerline of the aorta and the deployment of the stents is axisymmetric. Results of Stage 1 (III.C) show that these assumptions are a source of uncertainty during stent reconstruction, which may prevent in some cases to correctly simulate stent deployment. Additional steps of individual stent modeling and deployment corrections are therefore required.

2.4 Stage 2 (Refining Stage)

Individual stent deployment may not be accurate enough and need to be improved. This stage combines two individual refining steps: minimization of rotation and individual deployment of stents. The goal of the first step is to determine the actual angle of rotation Φ of each stent, around its main axis z'. The aim of the second step is to individually simulate the deployment of the stent in a deployment box. Both have clinical applications. When the SG is deployed, the actual SG deployment can be reconstructed in 3D. When the SG is not fully deployed, the EVAR practitioner can visualize how the SG would deploy at its current position.

2.4.1 Minimization of the Rotational Difference

During deployment, the stent may be subject to rotations Φ around its main axis $\mathbf{z}'$. In the case of axisymmetric stents, this rotation has little impact on its final deployment, although the surgeon may wish to improve the accuracy of reconstruction for critical stents. However, in the case of stents with fenestrations, their positioning depends on the rotation of the stent. It is therefore essential to determine these rotations. This step is performed within a minimization loop with a differential evolution algorithm.

In the case of axisymmetric stents, the value to be minimized is the difference e_S. This difference is calculated in two dimensions. The 3D model of the stent geometrically reconstructed at the end of the previous step is projected according to the projection parameters of the target image. e_S is the average distance between each point of the reconstructed stent and its nearest neighbor. This loop is used to determine the proper rotation $\Phi \pm k\theta$, where θ is the periodic angle separating two peaks of the Z-shape axisymmetric stent and k a real integer. Considering fenestrated SG, all stents with fenestrations or scallops have radiopaque markers to guide the positioning. The new e_M deviation to be minimized is calculated by considering only the distance between the radiopaque markers. In this case, the proper rotation Φ is exact and does not depend on θ. Indeed, the positioning of fenestrations is asymmetrical.

2.4.2 Individual Stent Deployment

The objective of this deployment box is to make maximum use of intraoperative image data. Thus, the deployment of the stent will be constrained not only by the geometry of the aorta, but also by the information from its convex hull on the image. From the convex hull obtained previously, we define a back-projection polyhedron (BpP). Each side of the BpP is a triangular element. The X-ray source is at the origin of the global coordinate system. Therefore, all triangular elements have a common point $\mathbf{O}$ $(0, 0, 0)$. Each edge of the convex hull is defined as the edge opposite to the apex $\mathbf{O}$ (Fig. 9). The volume is closed by the surface of the convex hull discretized in triangular elements. The Boolean intersection of the BpP with the aorta gives a volume called the deployment box. Boundaries of the volume are meshed with rigid shell elements. The stent is positioned in this rigid box according to its previously determined configuration.

The stents are composed of corotational Euler-Bernoulli beam elements. The mesh size is refined in high curvature areas. The stents are modelled in their 3D position and orientation determined in the previous steps. Only the diameter of the stent is changed. The stent model is initialized in its deployed configuration. Then it is pre-constrained to the diameter of the SG launcher before the simulation begins. Euler-Bernoulli beam elements have an elastic linear behaviour. The stents have a diameter of 0.125 mm and are made of 316 L steel which mechanical characteristics are summarized in [38].

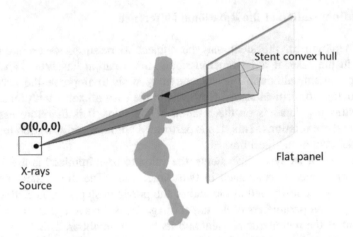

Fig. 9 Deployment box extracted from the intersection of the aorta volume and the BpP

Each stent, or a selection of stents at the practitioner's discretion, is deployed individually. The crimped stent is positioned inside the deployment box. The position of its barycenter and its orientation have already been recovered during the global positioning step. Then the stent is deployed (elastic recoil as the stent was crimped). The deployment is calculated using the Project Chrono engine [45], with solver Math Kernel Library (MKL) from Intel®. After the first contact, the time step is reduced to ensure the stability of the model. The contacts are modelled using the penalty algorithm implemented in Project Chrono, the Smooth-Contact (SMC) modeling approach. SMC uses penalty (in a discrete element method (DEM) [47, 48], regularizing the frictional contact forces, with "imaginary" spring-dashpot systems at each contact) and as such objects in contact will have slight interpenetration and integration time-step will likely be small. The simulations are performed on a computer with 4 CPUs, 3.40GHz, 16 GB RAM, but without parallelization. The computation time for deploying a stent is less than 6 minutes, without optimization. In addition, the complete calculation is easily parallelizable, as each stent deployment can be simulated on a separate core. The overall calculation time could therefore be compatible with clinical use.

3 Proof of Concept

The method described above was applied to a patient who underwent an EVAR procedure.

Table 1 Clinical summary of EVAR procedure

Sex	Male
Age (y)	78
Stent-graft:	ENBF-28-20-C-170-EE
Number of stents	20
Proximal graft diameter (mm)	28
Distal graft diameter (mm)	20
Anevrismal sac thrombus:	No

3.1 Clinical Data

Details about the device are given in Table 1. An additional stent was tested, including three fenestrations: right renal artery, mesenteric artery and left renal artery fenestrations. The preoperative and postoperative CT scans were acquired at the Saint-Etienne University Hospital under clinical conditions. Use of the clinical data was approved by the institutional review board and informed consent was obtained from the patient. The voxel size of the scans was $0.9395 \times 0.9395 \times 2$ mm^3.

First, we evaluated the results of Stage 1. As the results were not precise enough in terms of radial expansion, a second step of individual stent deployment was required. We evaluated the results of the complete method, including Correction Part. The average diameter of the renal arteries was 5–7 mm. If the fenestration positioning error is less than 3 mm, the surgeon can catheterize the secondary arteries such as renal arteries. Above this threshold, it is considered that intraoperative complications are likely to arise. Therefore, the clinical validation value was set at 3 mm in accordance with experienced clinicians.

3.2 Quality Assessment of Stent Deployment

First, to isolate and test the stent reconstruction algorithm as precisely as possible, the following assumptions were made: the projection matrix was known; the 3D geometry of the aorta was assumed to be perfectly registered. This ensured that not introducing positioning errors related to aortic registration. Thus, target images and the 3D model of the aorta were generated from the postoperative scan. The actual 3D position of the SG was therefore known, and served as a reference to be compared with the simulation for the sake of validation (Fig. 10).

Several parameters were used to assess the quality of the reconstruction. The first parameter was the D_B distance, which was the distance between the 3D barycentres of the target stent and the reconstructed stent. This distance was used to assess the quality of stent positioning within the artery. The second parameter was the distance D_M, which was the average distance between the point clouds of the target stent and the reconstructed stent. It was defined as the average of the Euclidean distances

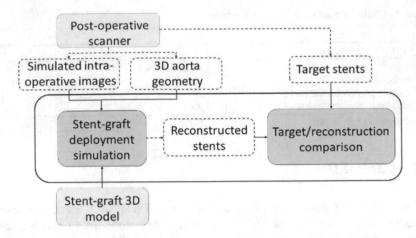

Fig. 10 Flowchart of the validation scheme

between a node of the reconstructed stent and its nearest neighbor among the points
of the target stent. This distance allowed reaching the quality of stent positioning
and deployment to be assessed at the same time. Finally, the last parameter is *CSAS*,
i.e. the cross-sectional area overlap of target and reconstructed stents, which allows
the quantitative comparison of stent deformations at the SGs folds. Cross-section
areas along the entire stent were measured after deployment in the aneurysm. After
deployment, the cross-sectional areas were modified, particularly in terms of SG
folds. Let A_T be the area of the target cross-section S_T and A_R the area of the
reconstructed cross-section S_R. A_U is the area of the S_U intersection between S_T
and S_R.

$$S_U = S_T \cup S_R \tag{9}$$

$$CSAS = 100 * \frac{A_T - |A_T - A_U|}{A_T} \ (\%) \tag{10}$$

3.3 Results and Discussion

Figure 11 shows the results of Stage 1 and Stage 2 and Table 2 shows a comparison
of the results of the two stages. Concerning Stage 1, the average D_B is generally
lower than the clinical validation value, the positioning of stents in the artery is
generally good. However, the error is too large on the contralateral limb, close to
or greater than 3 mm with a maximum of 3.71 mm. About D_M, the average error
is less than 3 mm, however the distance map shows that a significant number of
stents have an error too close to the limit. The *CSAS* map confirms this result, with

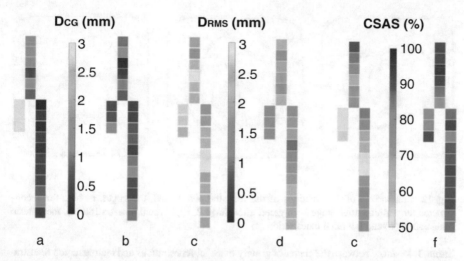

Fig. 11 Distance map representing the distance between the barycenters of the target and the reconstructed SG from Stage 1 (**a**) and Stage 2 (**b**), the RMS distance between the cloud points of the target and the reconstructed stents from Stage 1 (**c**) and Stage 2 (**d**), and the cross-sectional surface superposition between target and reconstructed stents from Stage 1 (**e**) and Stage 2 (**f**). Each square represents a stent

Table 2 Summary and comparison of Stage 1 and Stage 2 performances

		Preliminary part	Correction part	Difference	Difference (%)
D_B (mm)	Mean ± std	1.02 ± 1.03	0.75 ± 0.32	−0.27	−26.5
	Max	3.71	1.40	−2.31	−62.3
	Min	0.05	0.15	0,1	200.0
D_M (mm)	Mean ± std	2.21 ± 0.44	1.73 ± 0.37	−0.48	−21.7
	Max	3.12	2.28	−0.84	−26.9
	Min	1.54	0.90	−0.64	−41.6
CSAS (%)	Mean ± std	69.2 ± 12.8	86.6 ± 5.1	17.4	25.1
	Max	89.3	95.2	5.9	6.6
	Min	46.9	73.7	26.8	57.1

an average superposition of about 70% and a minimum of 47%, which is much too low. Thus, an additional step of individual deployment seems necessary. Stage 2 showed a clear improvement in the quality of stent simulation. All measurements were below the threshold value, with a maximum for the D_B of 1.40 mm and for the D_M of 2.28 mm. The mean values were also improved, by reducing the mean error of −26.5% for the D_B and − 21.7% for the D_M compared to Stage 1. There was also a clear improvement in the CSAS by a mean value of 25%.

Figure 12 shows how fenestration positions (front view) were predicted. The size difference between the target and reconstructed fenestrations results from the segmentation of the post-operative SG. The fenestrations are surrounded by radio-opaque markers that create artifacts during imaging. These artifacts make

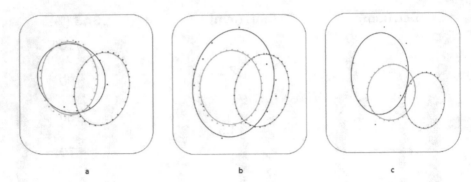

a   b   c

Fig. 12 Comparison of the positions of the fenestrations. In red, the target, in blue the reconstructed fenestration after Stage 1, in green after Stage 2. Right renal fenestration (**a**), mesenteric fenestration (**b**) and left renal fenestration (**c**)

Table 3 Distance between the center of gravity of target fenestrations and reconstructed fenestrations, before and after Stage 2

	Right renal fenestration	Mesenteric fenestration	Left renal fenestration
D_B Stage 1 (mm)	3.06	3.05	6.05
D_M Stage 2 (mm)	0.41	1.64	2.46

the markers appear larger than reality. The average distances between the centers of gravity of the fenestrations are summarized in Table 3. The overlapping of fenestrations is improved after the refining steps, which is visible on the Fig. 12. The method, after correction, was therefore able to position the fenestrations precisely enough to allow the catheterization of secondary arteries, with a maximum distance less than 2.46 mm for the left renal fenestration.

These results are interesting in comparison with previous work. Indeed, preoperative finite element simulations [21] have an average error of about 3.5 mm, which is higher than the results obtained here, with a longer computation time, as they are intended for planning purposes. It is though difficult to compare the previous fast methods with the one presented here. Indeed, the evaluation criteria are not similar. For example, in [33], the results of the fast method (FM) were compared with finite element simulations (FE), which may themselves differ from reality. In complex cases, the average distance between FM and FE is about 3–4 mm, for a very short calculation time (<1 min). However, these methods are not guided by intraoperative imaging. In [36], an accuracy of about 2–3 mm is reached, which is comparable to our results, with a shorter computation time. However, their method was based on additional markers bonded onto the device, which does not seem suitable. Compared to these previous studies, our method seems to achieve a promising compromise between accuracy, computation time and compatibility with current clinical conditions.

With a computer powered by 4 CPUs, 3.40GHz, 16 GB RAM, the calculation time for Stage 1 is under 20 seconds. The calculation time for the rotation minimization is under 1 minute. The maximum calculation time for the deployment simulation is 6 minutes. Co optimization and parallelization are currently ongoing to enable applications in clinical conditions.

The results of Stage 1 (first and second steps) show that the SG position is globally well predicted in the artery, but that the radial deployment of stents is not sufficiently accurate, with errors exceeding the 3 mm threshold. However, corrections made in Stage 2 (third and fourth step) yield a reasonable accuracy. More specifically, accurate predictions of fenestration positions after Stage 2 would avoid complex catheterization of secondary arteries, which is one of the major source of complications for practitioners.

However, the method has several limitations. First, it depends on the quality of the input data. Indeed, we assume that stents can be individualized using image processing. If this assumption is usually satisfied, overlapped stent can challenge it, for example, in the case of very pronounced angulation of the proximal neck of the aneurysm. In addition, the geometry of the aorta is assumed to be known, as it was updated using a non-rigid registration method previously. However, registration errors could occur and add to the other errors mentioned in the chapter. The non-rigid registration step also involves restriction for clinical applications. As we assume that we know the geometry of the aorta before the SG deployment simulation, the non-rigid registration must be performed each time the mobile C-arm position is changed, which implies to perform a DSA in order to visualize the aorta. The method should therefore only be used at key points in the procedure. Moreover, segmentation errors of the aorta volume are possible, but they are difficult to identify and probably have a minor influence on the results. A priori, the presence of thrombus is not a major problem for segmentation. Indeed, the aortic lumen is segmented during preoperative planning, not the aorta itself, and the SG is deployed in the lumen. Effects of thrombus on overall aortic stiffness are considered in the previous non-rigid registration step. But local variations in stiffness, due to thrombus, calcifications or surrounding tissues are not taken into account. Other limitations are related to the method itself. In order to save calculation time, the deployment of stents was simulated individually within rigid boxes. The rigid nature of the boxes is obviously a simplified mechanical behavior of the arterial wall but it did not induce significant errors when the simulations were compared with the post-operative CT scan. Finally, by simulating the individual deployment of stents, we do not consider the effect of the textile graft connecting them together. For example, a highly crimped stent can change the deployment diameter of the neighboring stent independently of the local geometry of the aorta. However, initial results show that these limitations do not hamper the accuracy more than what is compatible with clinical expectations. The influence of other assumptions should be investigated further with additional patient data.

This method would be particularly suited for complex cases of thoracoabdominal aortic aneurysms that require the use of fenestrated stent grafts. Indeed, when positioning fenestrations, the surgeon can potentially encounter difficulties. In most cases, the method is able to provide assistance to the surgeon. A few exceptions can challenge the method. The method is based on data extracted in the plane of the intraoperative image. Missing information along the projection axis are provided by simulations. If the main deployment axis of the stent graft was located along the projection axis, the method may have difficulty simulating the device. In practice, this situation seems very unlikely though.

4 Conclusion

We have presented in this chapter a methodology for simulating SG deployment based on intraoperative images coupled with mechanical models. The algorithm consists of a series of successive steps with increasing precision. A first step simulates the positioning of the SG within the aorta using a simplified finite element model and the centerline of the artery. The second step is an axisymmetric geometric reconstruction of the stents. The third step minimizes the rotation of the stent around its main axis. Finally, the last step consists in deploying each stent individually within a deployment box extracted from the imaging and geometry of the aorta.

The results of combined Stage 1 and Stage 2 yield a reasonable accuracy. More specifically, accurate positioning of fenestration would facilitate catheterization of secondary arteries. But the method suffers from several limitations. First, it depends on the quality of the input data and the ability of image processing to distinguish stents. In addition, the geometry of the aorta is supposed to be known. Next, the wall of the artery is considered rigid when the stents are deployed. Lastly, textiles are not simulated. The influence of these limitations on the accuracy of simulations needs to be explored further using additional patient data.

Finally, the translation of our methodology seems promising and could be generalized to all operating rooms equipped with mobile C-arms to assist SG deployment using real-time simulations in the future.

References

1. A. Dua, S. Kuy, C.J. Lee, G.R. Upchurch, S.S. Desai, Epidemiology of aortic aneurysm repair in the United States from 2000 to 2010. J. Vasc. Surg. **59**(6), 1512–1517 (2014)
2. M.S. Sajid, M. Desai, Z. Haider, D.M. Baker, G. Hamilton, Endovascular Aortic Aneurysm Repair (EVAR) has significantly lower perioperative mortality in comparison to open repair: A systematic review. Asian J. Surg. **31**(3), 119–123 (2008)
3. R.M. Greenhalgh, L.C. Brown, J.T. Powell, S.G. Thompson, D. Epstein, Endovascular repair of aortic aneurysm in patients physically ineligible for open repair. N. Engl. J. Med. **362**, 1872 (2010)
4. S. Macdonald, R. Lee, R. Williams, G. Stansby, Towards safer carotid artery stenting: A scoring system for anatomic suitability. Stroke **40**, 1698 (2009)
5. J.-N. Albertini et al., Aorfix stent graft for abdominal aortic aneurysms reduces the risk of proximal type 1 endoleak in angulated necks: Bench-Test study. Vascular **13**(6), 321–326 (2005)

6. J.-N. Albertini, J. Macierewicz, S. Yusuf, P. Wenham, B. Hopkinson, Pathophysiology of proximal perigraft endoleak following endovascular repair of abdominal aortic aneurysms: A study using a flow model. Eur. J. Vasc. Endovasc. Surg. **22**(1), 53–56 (2001)
7. A. Carroccio et al., Predicting iliac limb occlusions after bifurcated aortic stent grafting: Anatomic and device-related causes. J. Vasc. Surg. **36**(4), 679–684 (2002)
8. F. Cochennec, J.P. Becquemin, P. Desgranges, E. Allaire, H. Kobeiter, F. Roudot-Thoraval, Limb graft occlusion following EVAR: Clinical pattern, outcomes and predictive factors of occurrence. Eur. J. Vasc. Endovasc. Surg. **34**(1), 59–65 (2007)
9. S. Akpek, T. Brunner, G. Benndorf, C. Strother, Three-dimensional imaging and cone beam volume CT in C-arm angiography with flat panel detector. Diagn. Interv. Radiol. **11**, 10 (2005)
10. T. Fagan, J. Kay, J. Carroll, A. Neubauer, 3-D guidance of complex pulmonary artery stent placement using reconstructed rotational angiography with live overlay. Catheter. Cardiovasc. Interv. **79**(3), 414–421 (2012)
11. S.A.M. Baert, E.B. Van de Kraats, T. Van Walsum, M.A. Viergever, W.J. Niessen, Three-dimensional guide-wire reconstruction from biplane image sequences for integrated display in 3-D vasculature. IEEE Trans. Med. Imaging **22**, 1252 (2003)
12. M. Hoffmann et al., Semi-automatic catheter reconstruction from two views, in *Lecture Notes in Computer Science (Including Subseries Lecture Notes in Artificial Intelligence and Lecture Notes in Bioinformatics)*, Medical Image Computing and Computer-Assisted Intervention – MICCAI 2012. Lecture Notes in Computer Science, **7511**, (Springer, Berlin, Heidelberg, 2012)
13. M. Hoffmann et al., Reconstruction method for curvilinear structures from two views, in *Proceedings Volume 8671, Medical Imaging 2013: Image-Guided Procedures, Robotic Interventions, and Modeling; 86712F, Event: SPIE Medical Imaging*, (United States, Lake Buena Vista (Orlando Area), Florida, 2013)
14. M. Wagner, S. Schafer, C. Strother, C. Mistretta, 4D interventional device reconstruction from biplane fluoroscopy. Med. Phys. **43**(3), 1324–1334 (2016)
15. P. Mortier et al., A novel simulation strategy for stent insertion and deployment in curved coronary bifurcations: Comparison of three drug-eluting stents. Ann. Biomed. Eng. **38**(1), 88–99 (2010)
16. G.A. Holzapfel, M. Stadler, T.C. Gasser, Changes in the mechanical environment of stenotic arteries during interaction with stents: Computational assessment of parametric stent designs. J. Biomech. Eng. **127**(1), 166 (2005)
17. F. Auricchio, M. Conti, M. De Beule, G. De Santis, B. Verhegghe, Carotid artery stenting simulation: From patient-specific images to finite element analysis. Med. Eng. Phys. **33**(3), 281–289 (2011)
18. S. De Bock et al., Filling the void: A coalescent numerical and experimental technique to determine aortic stent graft mechanics. J. Biomech. **46**(14), 2477–2482 (2013)
19. C. Kleinstreuer, Z. Li, M.A. Farber, Fluid-structure interaction analyses of stented abdominal aortic aneurysms. Annu. Rev. Biomed. Eng. **9**(1), 169–204 (2007)
20. F. Auricchio, M. Conti, S. Marconi, A. Reali, J.L. Tolenaar, S. Trimarchi, Patient-specific aortic endografting simulation: From diagnosis to prediction. Comput. Biol. Med. **43**(4), 386–394 (2013)
21. D. Perrin et al., Patient-specific numerical simulation of stent-graft deployment: Validation on three clinical cases. J. Biomech. **48**(10), 1868–1875 (2015)
22. D. Perrin et al., Deployment of stent grafts in curved aneurysmal arteries: Toward a predictive numerical tool. Int. J. Numer. Method. Biomed. Eng. **31**(1), e02698 (2015)
23. D. Perrin et al., Patient-specific simulation of stent-graft deployment within an abdominal aortic aneurysm, in *CMBE Proceedings Series, 3rd International Conference on Computational and Mathematical Biomedical Engineering, Hong Kong*, 16–18, Dec 2013
24. D. Perrin et al., Patient-specific simulation of endovascular repair surgery with tortuous aneurysms requiring flexible stent-grafts. J. Mech. Behav. Biomed. Mater. **63**, 86–99 (2016)
25. S. Appanaboyina, F. Mut, R. Lohner, C.M. Putman, J.R. Cebral, Computational fluid dynamics of stented intracranial aneurysms using adaptive embedded unstructured grids. Int. J. Numer. Methods Fluids **57**, 475 (2008)

26. L. Flórez Valencia, J. Montagnat, M. Orkisz, 3D models for vascular lumen segmentation in MRA images and for artery-stenting simulation. IRBM **28**(2), 65–71 (2007)
27. L. Flórez-valencia et al., 3D graphical models for vascular-stent pose simulation to cite this version. HAL id: HAL-00682926. Innovations Technol. Biol. Med. **13**(3), 235–248 (2007)
28. I. Larrabide, A. Radaelli, A. Frangi, *Fast Virtual Stenting with Deformable Meshes: Application to Intracranial Aneurysms* (Springer, Berlin, Heidelberg, 2008), pp. 790–797
29. I. Larrabide, M. Kim, L. Augsburger, M.C. Villa-Uriol, D. Rüfenacht, A.F. Frangi, Fast virtual deployment of self-expandable stents: Method and in vitro evaluation for intracranial aneurysmal stenting. Med. Image Anal. **16**(3), 721–730 (2010)
30. E. Flore, I. Larrabide, L. Petrini, G. Pennati, A. Frangi, Stent deployment in aneurysmatic cerebral vessels: Assessment and quantification of the differences between Fast Virtual Stenting and Finite Element Analysis, CI2BM09 – MICCAI Workshop on Cardiovascular Interventional Imaging and Biophysical Modelling, Not in a journal, conference proceedings, Sep. 2009
31. A. Bernardini et al., Deployment of self-expandable stents in aneurysmatic cerebral vessels: Comparison of different computational approaches for interventional planning. Comput. Methods Biomech. Biomed. Engin. **15**(3), 303–311 (2012)
32. K. Spranger, Y. Ventikos, Which spring is the best? Comparison of methods for virtual stenting. IEEE Trans. Biomed. Eng. **61**(7), 1998–2010 (2014)
33. K. Spranger, C. Capelli, G.M. Bosi, S. Schievano, Y. Ventikos, Comparison and calibration of a real-time virtual stenting algorithm using Finite Element Analysis and Genetic Algorithms. Comput. Methods Appl. Mech. Eng. **293**, 462–480 (Aug. 2015)
34. J. Zhong et al., Fast virtual stenting with active contour models in intracranial aneurysm. Sci. Rep. **6**(January), 1–9 (2016)
35. S. Demirci et al., 3D stent recovery from one x-ray projection. Lect. Notes Comput. Sci. **6891 LNCS**(PART 1), 178–185 (2011)
36. X.-Y. Zhou, J. Lin, C. Riga, G.-Z. Yang, S.-L. Lee, Real-time 3-D shape instantiation from single fluoroscopy projection for fenestrated stent graft deployment. IEEE Robot. Autom. Lett. **3**(2), 1314–1321 (Apr. 2018)
37. J.-Q. Zheng, X.-Y. Zhou, C. Riga, G.-Z. Yang, Real-time 3D shape instantiation for partially-deployed stent segment from a single 2D fluoroscopic image in fenestrated endovascular aortic repair. IEEE Robot. Autom. Lett., 1–1 (2019)
38. N. Demanget et al., Computational comparison of the bending behavior of aortic stent-grafts. J. Mech. Behav. Biomed. Mater. **5**(1), 272–282 (2012)
39. A.F. Frangi, W.J. Niessen, K.L. Vincken, M.A. Viergever, *Multiscale vessel enhancement filtering* (Springer, Berlin, Heidelberg, 1998), pp. 130–137
40. The Vascular Modeling Toolkit, 2019. [Online]. Available: www.vmtk.org
41. M. Groher, D. Zikic, N. Navab, Deformable 2D-3D registration of vascular structures in a one view scenario supplementary material derivative of the difference measure. IEEE Trans. Med. Imaging **28**(6), 847–860 (2009)
42. J.-Q. Zheng, X.-Y. Zhou, C. Riga, G.-Z. Yang,. 3D Path Planning from a Single 2D Fluoroscopic Image for Robot Assisted Fenestrated Endovascular Aortic Repair: arXiv : arXiv preprint arXiv:1809.05955
43. A. Pionteck, *Finite-Element Based Image Registration for Endovascular Aortic Aneurysm Repair. PhD Thesis, Mines Saint-Etienne, University of Lyon* (Saint-Etienne, France, 2020)
44. D. T. Project Chrono, *Chrono: An Open Source Framework for the Physics-Based Simulation of Dynamic Systems*. [Online]. Available: https://github.com/projectchrono/chrono
45. A. Tasora et al., Chrono: An open source multi-physics dynamics engine, in *Lecture Notes in Computer Science (Including Subseries Lecture Notes in Artificial Intelligence and Lecture Notes in Bioinformatics)*, High Performance Computing in Science and Engineering. HPCSE 2015. Lecture Notes in Computer Science, Springer, Cham **9611**, 19–49 (2016)
46. A. Tasora., *Euler-Bernoulli Corotational Beams in Chrono :Engine technical documentation*, http://www.projectchrono.org/. 1–12 (2016)

47. A. Tasora, M. Anitescu, S. Negrini, D. Negrut, A compliant visco-plastic particle contact model based on differential variational inequalities. Int. J. Non. Linear. Mech. **53**, 2–12 (2013)
48. P.A. Cundall, O.D.L. Strack, A discrete numerical model for granular assemblies. Géotechnique **29**(1), 47–65 (1979)

Efficient GPU-Based Numerical Simulation of Cryoablation of the Kidney

Joachim Georgii, Torben Pätz, Christian Rieder, Hanne Ballhausen,
Michael Schwenke ⓘ, Lars Ole Schwen ⓘ, Sabrina Haase,
and Tobias Preusser

Abstract Cryoablation, a minimally invasive technique for treating cancer, could benefit from computer support in planning, intervention and follow-up. For employing such treatment planning in daily clinical routine, individualized simulation of cryoablation needs to be sufficiently accurate and fast. This paper describes a simulation of cryoablation of human kidney permitting high-performance simulations on graphics hardware. The simulation involves partial differential equations modeling temperature evolution and phase changes in the tissue, as well as equations describing the dependence of tissue parameters on tissue temperature. A mushy region approach and a predictor-corrector time stepping scheme are utilized for discretization to achieve an efficient numerical scheme implemented on graphics hardware. The simulation is planned to be integrated in an approved medical device.

Keywords Cryoablation · Enthalpy approach · Tissue parameter model · Numerical simulation · GPU computing

1 Introduction

Thermal ablation for cancer treatment has reached significant attention in recent decades. As a minimally invasive technique, it allows treating patients with palliative and curative intent when surgery is not possible. In hyper-thermal ablation, energy is introduced into the tissue, heating it locally up to temperatures at which

---

J. Georgii and T. Pätz have contributed equally to this work.

J. Georgii · T. Pätz · C. Rieder · H. Ballhausen · M. Schwenke · L. O. Schwen · S. Haase
Fraunhofer Institute for Digital Medicine MEVIS, Bremen, Germany

T. Preusser (✉)
Fraunhofer Institute for Digital Medicine MEVIS, Bremen, Germany

Jacobs University, Bremen, Germany
e-mail: tobias.preusser@mevis.fraunhofer.de

© Springer Nature Switzerland AG 2020
K. Miller et al. (eds.), *Computational Biomechanics for Medicine*,
https://doi.org/10.1007/978-3-030-42428-2_11

proteins denaturate and cells die. The energy entry is achieved through laser light (LITT), microwaves (MWA), radio-frequency current (RFA), or high intensity focused ultrasound (HIFU, FUS) [8]. In hypo-thermal percutaneous ablation (cryoablation), the focus of this paper, an applicator (cryoprobe) is inserted into the target structure, cooling the tissue below freezing. The general treatment protocols include an alternating sequence of cooling, passive thawing (no cooling), and active thawing (warming). During freezing and thawing, complex processes take place in the tissue, including crystallization of water, cell dehydration, metabolic derangement, and vascular stasis [9]. Cyroablation has been in clinical use for many decades. Today it is performed in a variety of organs [34], including prostate, kidney, bone, breast, and pancreas.

For the aforementioned ablation techniques, it has been hypothesized that computer support for the planning, execution, and follow-up allows achieving better outcome [7, 15, 27]. In this context, many research activities on mathematical modeling, numerical simulation, and optimization of thermal ablation have been conducted in the past decades. In addition to tasks of image processing, such as registration, segmentation, and quantification, computer models are used to simulate the temperature fields or tissue destruction on patient-specific image data. Based on such thermal simulation, optimal probe placement can also be calculated by computer programs. For post-treatment follow-up, image registration can be used to assess the result of the treatment.

Clinical applicability of numerical simulation and optimization, however, faces several constraints and challenges:

1. Patient-specific data on biophysical tissue properties is not available
2. Computational power is mostly limited to contemporary desktop/laptop PCs
3. Established clinical workflows only allow for small timeframes in which a numerical simulation can take place.

Thus, any effort on numerical simulation or optimization intending to yield an impact in the clinical setting must strike a balance in the "force-field" created by these constraints: Computational accuracy must be balanced with limited computational power, uncertainty of tissue data must be quantified, and sensitivity of the simulation results with respect to data uncertainty must be analyzed.

In this paper, we report our activities in the development of a high-performance simulation of cryoablation that runs on contemporary graphics hardware and is fast enough to be utilized in the daily clinical routine. Our work merges findings of previous authors on modeling cryoablation and temperature-dependent tissue parameters (see citations below and in the following sections) in an engineering effort. We leave open scientific questions untouched and instead choose a pragmatic approach towards a numerical simulation with appropriate accuracy and speed. The cryoablation simulation described in this paper has been quality assured and developed according to ISO13485 [13]. Together with an industry partner, the simulation has been validated and is planned to be integrated into an approved medical device.

Our simulation is to be utilized by a user (clinical doctor) with access to patient-specific image data (CT or MRI) of the patient's kidney. Through image segmentation, the user must have identified the organ, the target region (lesion or tumor), blood vessels, and risk structures. The segmented image regions determine tissue classes to be considered as an input to the numerical simulation. Based on parameters of a cryoprobe and a user-prescribed cryoablation protocol, i.e., a cycle of alternating phases of freezing and thawing, the simulation calculates the temperature field at the end of the ablation cycle. The user can decide to overlay this temperature field with the segmented image data and draw conclusions about possible tissue damage.

Related Work In a related approach, Rossi, Rabin et al. [22, 23] presented efficient numerical schemes and approaches to experimental validation of cryosurgery. Keelan et al. [14] investigated GPU-based bioheat simulation in the context of cryoablation and its use as a training tool. Further, Rabin et al. [19, 28] presented a training tool for prostate cryosurgery. Zhang et al. [35] discussed numerical simulation of cryoablation in prostate cancer. Furthermore, Baissalov et al. [2] worked on an in-silico treatment planning used for optimization of multi-probe cryoablation. We will cite further publications throughout this paper in the realm of modeling of cryosurgery and the dependence of material parameters on tissue temperature.

Structure of the Paper The mathematical model for cryoablation is presented in the following Sect. 2. A tissue parameter model is discussed in Sect. 3. Thereafter, Sect. 4 presents our discretization tailored towards implementation on GPU hardware as described in Sect. 5. Some numerical tests and an example using real patient data are shown in Sect. 6. Finally, we summarize our work, draw conclusions and give an outlook on future work in Sect. 7.

2   Mathematical Model for Cryoablation

Our goal is to achieve a simulation of the tissue temperature T and the tissue freezing state F in the 3D spatial domain $\Omega = \Omega_{\text{tissue}} \cup \Omega_{\text{needle}} \cup \Omega_{\text{active}} \subset I\!R^3$ and within the time interval $I := [0, t_{\text{end}}]$ where t_{end} denotes the overall duration of a freeze-thaw cycle of a cryoablation of the kidney. Here, $\Omega_{\text{needle}} \cup \Omega_{\text{active}}$ denotes the volume covered by the ablation probe(s), Ω_{active} is their active zone(s), and Ω_{tissue} the surrounding tissue. We refer to Fig. 1 for a schematic illustration of these domains in 2D. Note that Ω_{needle} and Ω_{active} may denote the respective volumes of multiple needles that are used simultaneously and are driven by the same therapy protocol. We assume, however, that the isolated needle parts have negligible influence on the temperature evolution and let $\Omega_{\text{needle}} = \{\}$. The outer boundary of Ω is denoted by $\partial\Omega$.

Fig. 1 Computational
domain sketched in 2D

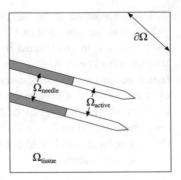

Fig. 2 Enthalpy H vs
temperature T during phase
transition between liquid and
frozen water

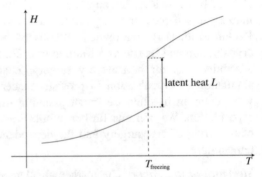

2.1 Temperature Model

Modeling temperature during cryoablation involves the phase change from liquid water to frozen water. Thus, the governing equations must consider enthalpy per volume $H = \rho c T$, i.e., thermal energy and latent heat L, the energy released in the freezing process. At the freezing point T_{freezing}, the enthalpy jumps by an amount equal to the latent heat, see Fig. 2. The heat conduction process is described by (cf. [12])

$$\frac{\partial}{\partial t} (\rho c T + L) = \nabla \cdot (k \nabla T) + Q \qquad \text{in } I \times \Omega, \tag{1}$$

with appropriate boundary and initial conditions, see below. Here and in the following, $T(t, x)$ denotes temperature (in K), $L(t, x)$ denotes latent heat (in J/m³), $\rho(t, x; T, F)$ denotes density (in kg/m³), $c(t, x; T, F)$ denotes specific heat capacity (in J/kg K) also called specific heat, and $k(t, x; T, F)$ denotes the heat conductivity (in W/mK). The variables ρ, c, and k are patient-specific tissue parameters whose modeling and the associated uncertainty are discussed below. We emphasize that ρ, c, and k depend on the temperature T and the freezing state F. Heat sources and sinks are denoted by $Q(t, x)$.

Latent Heat Our approach models a lumped energy balance which does not distinguish between intracellular and extracellular constituents of the tissue. Instead, the thermodynamical properties result from the volume fractions of ice and water. Thus, the latent heat L depends on the mass fraction of water present which is capable of undergoing freezing and thus can be expressed as

$$L(t, x) = l\rho(t, x; T, F)\Lambda(1 + F(T)) \qquad (2)$$

where l is the mass-specific latent heat (in J/kg) and Λ (unitless) is the total volume fraction of water within a unit volume of tissue (water content). The state $F \in [-1, 0]$ (unitless) represents the phase of water. A value of -1 indicates a fully frozen unit volume of water, and a value of 0 indicates a fully liquid unit volume of water. Thus, $1 + F$ is the mass fraction of liquid water.

Heat Sources and Sinks When modeling thermal ablation, the source term $Q(t, x)$ in (1) considers

(A) volumetric heat sources and sinks $q_s(t, x)$,
(B) the heat sink[1] effect by capillary perfusion $q_p(t, x)$ and larger blood vessels,
(C) the heat induced by metabolic activity $q_m(t, x)$.

A. Volumetric Heat Sources and Sinks The energetic interaction between the cryoablation needles and the surrounding tissue is purely through heat diffusion through the surfaces of the needles. Because the needle is kept at an extremely low temperature by the cryoablation device, we use a Dirichlet boundary condition rather than a volumetric source/sink term. Thus, we let $q_s(t, x) = 0$.

B. Capillary Perfusion We model the heat sink effect due to capillary perfusion through an effective conductivity term [33]. Thus, we let $q_p(t, x) = 0$ and we replace the thermal conductivity k with

$$k_{\text{eff}}(t, x; T, F) = k(t, x; T, F)(1 + \alpha w_{\text{blood}}(t, x; F)) \qquad (3)$$

(again in W/mK). The (unit-less) blood perfusion rate w_{blood} is known from the classical Pennes model [17]. The value $\alpha = 2$ (unitless) is based on Shih et al. [29] and describes the local blood vessel size and density.

C. Metabolic Activity Because metabolic activity ceases when temperatures approach freezing, it is safe to let $q_m(t, x) = 0$.

Blood Vessels In our approach, we model only capillary perfusion. Larger blood vessels could be modeled through distinct material properties or by an additional internal Dirichlet boundary. However, there is no bench data available that allows developing a model for the heat sink effect of blood vessels during cryoablation, and

---

[1]Note that in cryoablation, blood vessels act as "warming devices". Nevertheless, we employ the widely used "heat *sink* effect" notion seen in the literature.

the limited amount of available clinical data is not sufficient for a rigorous heat sink model. From clinical data, the heat sink effect of blood vessels during cryoablation is known to be less prominent than for hyperthermal ablation, i.e., the ablation area is close to ellipsoidal, and the effects of different tissue characteristics surrounding the needles seem to dominate the influence of the vessels in the available data. Thus, we refrain from modeling large blood vessels explicitly.

Combining all equations from above as the temperature model, we arrive at the following extension of the classical bioheat transfer equation

$$\frac{\partial}{\partial t}(\rho c T) = \nabla \cdot (k_{\text{eff}} \nabla T) - l \Lambda \frac{\partial}{\partial t}(\rho(1 + F)) \qquad \text{in } I \times \Omega. \tag{4}$$

Boundary and Initial Conditions Equation (4) is a parabolic PDE to be solved in the time-space cylinder $I \times \Omega$ with appropriate boundary and initial conditions: Because tissue is at body temperature ($T_{\text{body}} = 310.15\,\text{K}$) at the beginning of the therapy, a straightforward and meaningful initial condition is given by $T(0, x) = T_{\text{body}}$ and $F(0, x) = 0$ for $x \in \Omega$. Assuming that the computational domain Ω is large enough that the cooling of the cryoprobe does not interfere with the body regions outside of the computational domain, we consider the outer boundary $\partial\Omega$ to be a Dirichlet boundary with a constant body temperature, so $T(t, x) = T_{\text{body}}$ for $t \in I$ and $x \in \partial\Omega$.

The inner boundary $\partial\Omega_{\text{active}}$ is used during the freezing and active thawing cycle only. During passive thawing the temperature in Ω_{active} is not fixed, but follows the diffusion equation (4) like all other parts of the domain Ω. During active thawing, the temperature in Ω_{active} is set to T_{body}. Finally, during the freezing phases, we prescribe a fixed temperature T_{needle} on the active zones of the instrument.

In summary, the boundary conditions on the active zone are

$$T(t, x) = \begin{cases} T_{\text{needle}} & \text{if } x \in \partial\Omega_{\text{active}}, t \in \text{freezing phase,} \\ T_{\text{body}} & \text{if } x \in \partial\Omega_{\text{active}}, t \in \text{active thawing phase.} \end{cases} \tag{5}$$

2.2 Freezing Model

To model tissue freezing, we employ a mushy region approach in which $F(T)$ denotes the volume fraction of water that has undergone the phase change from liquid to frozen. We use an explicit relation between the temperature T and the freezing state F of the form

$$F(T) = \begin{cases} -1 & \text{if } T \leq T_0, \\ \dfrac{T_1 - T}{T_0 - T_1} & \text{if } T_0 < T < T_1, \\ 0 & \text{if } T \geq T_1. \end{cases} \tag{6}$$

We follow [20, 22] and choose $T_0 = 247.15\,\text{K}$ and $T_1 = 273.15\,\text{K}$. Note that by the chain rule, $\frac{\partial}{\partial t}F(T) = \frac{\partial F}{\partial T}\frac{\partial T}{\partial t} = F'\frac{\partial T}{\partial t}$, the derivative of F can be calculated piecewise in (6). Regarding the influence of enthalpy on the energy balance in Eq. (4), this result means that an additional change of temperature $F'\frac{\partial T}{\partial t}$ is active only during the phase transition, i.e., when $T \in (T_0, T_1)$.

With the mushy region approach, the jump in the enthalpy per volume at freezing is converted into an invertible approximation given by

$$
H(T) = \begin{cases}
\displaystyle\int_0^T \rho(\theta)c(\theta)\,d\theta & \text{if } T \leq T_0, \\[2ex]
\displaystyle\int_0^T \rho(\theta)c(\theta)\,d\theta + (1 + F(T))\,l\rho(T)\Lambda & \text{if } T_0 < T < T_1, \\[2ex]
\displaystyle\int_0^T \Big(\rho(\theta)c(\theta) + l\rho(\theta)\Lambda\Big)\,d\theta & \text{if } T \geq T_1.
\end{cases} \tag{7}
$$

2.3 Damage Model

Various approaches to model the freezing-induced tissue damage have been reported [9]. The criteria deduced for tissue damage vary significantly. Therefore, our simulation only provides the temperature distribution at the end of the simulated freeze/thaw cycle. An interpretation of this temperature distribution is left to a clinical expert that would be using the simulation. Consequently, all tissue parameters have to be modeled with dependency on the tissue temperature and not on the tissue state.

3 Tissue Parameter Model

The model equations from the previous sections involve tissue parameters that are state-dependent (cf. (1)) and spatially resolved. In our approach, we introduce a label function $\mathscr{T} : \Omega \rightarrow \mathscr{L}$. Each spatial location x in the computational domain Ω is assigned one of the tissue types in the set of tissue labels $\mathscr{L} = $ {kidney, kidney tumor, kidney background}. The tissue label *kidney background* represents the mixture of different tissues outside the organ of interest. These may appear in the computational domain when it does not completely lie inside the organ but crosses its boundaries. When using the simulation for patient specific treatment planning, these domains need to be obtained by segmentation of the corresponding patient specific image data.

We assume that all parameters depend on the type of tissue $\mathscr{T}$ and its temperature T. Furthermore, the heat capacity c and thermal conductivity k depend on the freezing state F. However, due to the coupling of the freezing state and the temperature via the mushy region approach (cf. (6)), this dependency can be expressed solely through the temperature T. Thus, our tissue parameters are of the form $c(t, x; T, F) = c(\mathcal{T}(x), T(t, x))$ etc. The effective thermal conductivity k_{eff} results, as before, from (3).

We maintain the density ρ with respect to changes in the temperature T to account for mass conservation in our model equations and their discretization, i.e., $\rho(t, x; T, F) = \rho(\mathcal{T}(x))$.

Thermal Conductivity k As described above, k is given in units W/mK. We omit units for constants throughout this section to simplify notation.

Kidney Pham and Willix [18] investigated the change of thermal conductivity in ex-vivo thermodynamical measurements of fresh lamb meat and suggest

$$k(T) = a + bT + \frac{c}{T} \qquad \text{for } T < T_{\text{freezing}} \tag{8}$$

for appropriately chosen values of a, b, and c. In this equation, the term bT accounts for the variation of thermal conductivity of ice with temperature. The term c/T models the change of thermal conductivity with the fraction of frozen water, which itself varies as $1 - \frac{T_{\text{freezing}}}{T}$. Pham and Willix set $T_{\text{freezing}} = 272.25$ K ($-1°$C) as an approximation to the temperature at which the jump occurs in the enthalpy-temperature curve, see Fig. 2. For temperatures above freezing, i.e., $T > T_{\text{freezing}}$, the c/T term is omitted, thus

$$k(T) = d + eT \tag{9}$$

with appropriate values d and e.

Pham and Willix used linear regression and interpolation to find the corresponding values of $a, \ldots, e$ from their measurement data. We use their results for lamb kidney tissue and reformulate equations (8) and (9) to arrive at

$$k(\text{kidney}, T) = \begin{cases} \begin{aligned} & 0.507 - 0.0075(T - T_{\text{freezing}}) \\ & +0.78\left(\frac{1}{(T - 273.15)} - \frac{1}{(T_{\text{freezing}} - 273.15)}\right) \end{aligned} & \text{if } T \leq T_{\text{freezing}}, \\[2em] 0.507 + 0.0012(T - T_{\text{freezing}}) & \text{if } T > T_{\text{freezing}}. \end{cases} \tag{10}$$

Kidney Tumor In [31], tumor tissue is reported to show significantly different properties than water. Deshazera et al. [6] refer to studies that show that the thermal conductivity of tumors resected from humans and animals is as much as 20% higher than healthy liver tissue. We apply this finding for kidney tumors and set $k(\text{kidney tumor}, T) = 1.2 \, k_{\text{kidney}}(T)$.

Kidney Background We assume that fat tissue dominates in the tissues that surround the liver. To the authors knowledge, no investigations on the temperature dependence of the thermal conductivity of human fat tissue have been reported in the literature. ITIS [11] reports values at body temperature of 310.15 K.

Consequently, we base our model on the work of [10], who investigated the thermal conductivity of dairy products, whose fat, like human fat tissue, mainly consists of triglycerides. They report affine linear dependence, i.e., $k(T) = a + bT$. Empirical equations are given for different fat content, e.g., for 60% and 80% as

$$k_{\text{cream, 60\% fat}}(T) = 0.1743 + 0.001264 \cdot (T - 273.15),$$
$$k_{\text{cream, 80\% fat}}(T) = 0.1653 + 0.000997 \cdot (T - 273.15). \tag{11}$$

Assuming water content of 30% for adipose tissue [26] and 70% fat content, we derive the thermal conductivity for adipose tissue as mean value of these equations, thus

$$k(\text{fat}, T) = \frac{1}{2}\left(k_{\text{cream, 60\% fat}}(T) + k_{\text{cream, 80\% fat}}(T)\right). \tag{12}$$

For temperatures below the freezing, we directly adopt the values listed in Choi et al. [4] for bovine fat, see Table 1.

Specific Heat Capacity c

Kidney Choi and Bischoff [4] present values for the specific heat capacity in lamb kidney at different temperatures below freezing. Rossmann and Haemmerich [24] present values for temperatures above the freezing. We take the values from these references as shown in Table 2 and we interpolate linearly between them.

Table 1 Values of k in fat

T		k
[°C]	[K]	$[\frac{W}{mK}]$
−18	255.15	0.28
−9.4	263.75	0.3
−7.6	265.55	0.216
−5	268.15	0.266
0.1	273.25	0.193
37	310.15	0.21145

Table 2 Values of c in kidney

T		c	
[°C]	[K]	$[\frac{J}{kgK}]$	Reference
−40	233.15	1630	[4]
−23	250.15	2140	[4]
10	283.15	3505	[24]

Table 3 Values of c in fat

T		c	
[°C]	[K]	$[\frac{J}{kg\,K}]$	Reference
−160	113.15	865	[4]
−120	153.15	1015	[4]
−80	193.15	1380	[4]
−40	233.15	1950	[4]
20	293.15	2348	[11]

Table 4 Values of ρ

Tissue type	ρ $[\frac{kg}{m^3}]$	Comments & references
Kidney	1066	[11]
Kidney tumor	1066	Like native tissue
Fat (kidney bg)	911	[11]

Table 5 Values of Λ

Tissue type	Λ [%]	Comments & references
Kidney	79.47	[16]
Kidney tumor	83.44	Kidney + 5% [21]
Fat (kidney bg)	30	[26]

Kidney Tumor Due to lack of data, we pragmatically set c(kidney tumor, T) = c(kidney, T).

Kidney Background Again, Choi et al. [4] serves as our reference for values of the specific heat capacity of bovine and pig fat (which is the kidney background tissue) between 113.15 K and 233.15 K. Because the references give a range (min/max) of heat capacity values, we use the averages of porcine and bovine fat values for the available temperature samples as shown in Table 3.

Density ρ For reasons of energy and mass conservation, we assume that the tissue density does not change with temperature, see Table 4.

Water Content Λ We base the values for water content on [16] and [25]. Furthermore, according to Ross [21], tumor tissue in the liver has up to 5% more water than normal liver tissue. We also use this observation for the water content of kidney tumors, see Table 5.

Perfusion Rate w_{blood} In our model, the relative perfusion rate w_{blood} is needed as part of the heat sink term in the bioheat transfer Eq. (4). Thus, this parameter refers to capillary perfusion in the tissue in contrast to blood flow in larger blood vessels. Because the tissue freezes below $T_{freezing}$, we set the relative capillary perfusion rate w_{blood} to zero below this temperature. No information on the temperature dependence of w_{blood} can be found in the literature. Therefore, we interpolate linearly between $T_{freezing}$ and T_{body}, where the values of w_{blood} for T_{body} are computed from the specific perfusion rates reported in the literature with the density

Table 6 Values of spec. perfusion rate and resulting rel. perfusion rate w_{blood}

Tissue type	Spec. perf. rate [$\frac{m^3}{kg}\frac{1}{s}$]	w_{blood} [s^{-1}]	Comments & references
Kidney	$6.66 \cdot 10^{-5}$	0.071	[30]
Kidney tumor	$6.66 \cdot 10^{-5}$	0.071	Like native tissue
Fat (kidney bg)		0.00175	Estimated, [3, 30]

values from Table 4. Due to the lack of further information for higher temperatures and because we do not consider a warming process, we pursue with constant perfusion rate above body temperature in our approach. Due to the lack of data for tumor tissue types, we use the values of native tissue instead. The resulting values are reported in Table 6.

4 Discretization

4.1 Temporal Discretization

For increased performance on the proposed graphics hardware architecture (GPU computing), we choose an explicit time-stepping scheme with a fixed time step size τ. Equidistant time points $t_i = i\tau$ are introduced such that the final time point t_{end} is reached and such that the minimal number of time steps J is used with $t_{end} = J\tau$ and τ. The time step size τ is deliberately not chosen adaptively but such that it fulfils the CFL condition [5] asserting stability for all possible values of the material coefficients and their state dependency. The selection of the stable time step is detailed below.

In the following, we denote time discrete quantities with a superscript (i), i.e., for $i = 0, \ldots, J$ we write $T^{(i)}(x) = T(t_i, x)$, $F^{(i)}(x) = F(T^{(i)}(x))$, etc. For later use, we introduce the set of boundary points $\Gamma^{(i)}$, which varies according to the phases of the cryoablation cycle, thus,

$$\Gamma^{(i)} = \partial\Omega \cup \begin{cases} \partial\Omega_{active} & \text{if } t_i \in \text{freezing phase or active thawing phase,} \\ \emptyset & \text{else.} \end{cases} \tag{13}$$

Also, we introduce the temperature at the active tip of the cryoprobe for each time step in which the active zone acts as Dirichlet boundary as

$$T_{probe}^{(i)} = \begin{cases} T_{needle} & \text{if } t_i \in \text{freezing phase,} \\ T_{body} & \text{if } t_i \in \text{active thawing phase.} \end{cases} \tag{14}$$

Predictor Step for Temperature Computation As a central ingredient of our time-stepping scheme, we evaluate all nonlinearities at the old time step t_i, assuming that tissue parameters change slowly with temperature and freezing state and thus can be regarded as constant per time step. Consequently, we can use a simple forward difference quotient to approximate

$$\frac{\partial}{\partial t}(\rho c T)(t_i, x) \approx \rho^{(i)}(x) c^{(i)}(x) \frac{T^{(i+1)}(x) - T^{(i)}(x)}{\tau}. \tag{15}$$

In this approximation, we intentionally omit the time derivatives of ρ and c because we assume that they are constant per time step. Using this, a prediction of the updated temperature is obtained by

$$T^{(i+1/2)}(x) = T^{(i)}(x) + \frac{\tau}{\rho^{(i)}(x) c^{(i)}(x)} \left[\nabla \cdot \left(k_{\text{eff}}^{(i)}(x) \nabla T^{(i)}(x) \right) \right]. \tag{16}$$

Corrector Step for Temperature Computation To approximate the enthalpy term of (4), we use a correction step that allows us to achieve an enthalpy-conserving discrete scheme for temperatures within the mushy region $[T_0, T_1]$. To this end, we approximate the nonlinear enthalpy-temperature relation from Eq. (7) (see Fig. 2) by an idealized and invertible approximation $\tilde{H}^{(i)}$:

$$\tilde{H}^{(i)}(T) = \begin{cases} c^{(i)} T - l\Lambda & \text{if } T \le T_0, \\ c^{(i)} T + \left(\dfrac{T - T_0}{T_1 - T_0} - 1 \right) l\Lambda & \text{if } T_0 < T < T_1, \\ c^{(i)} T & \text{if } T \ge T_1. \end{cases} \tag{17}$$

This approximation $\tilde{H}$ interpolates the slope of the enthalpy H locally around $T^{(i)}$, see Fig. 3 (left). Based on Eq. (17), we evaluate the current enthalpy of the prediction step as

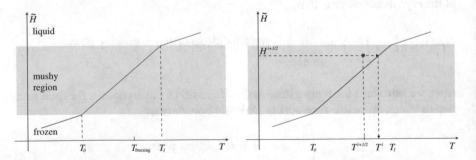

Fig. 3 *Left*: Enthalpy vs. temperature curve in an idealized setting used in the discretization. *Right*: The temperature correction step projects the temperature estimate $T^{(i+1/2)}$ back to the curve of $\tilde{H}$, keeping the local enthalpy $H^{(i+1/2)}$ constant

$$H^{(i+1/2)} = c^{(i)} T^{(i+1/2)} + \left(\frac{T^{(i)} - T_0}{T_1 - T_0} - 1 \right) l\Lambda. \tag{18}$$

We consider the freezing state from the old temperature $T^{(i)}$ to evaluate the latent heat. Due to the unrestricted temperature update from Eq. (15), the temperature/enthalpy pair $(T^{(i+1/2)}, H^{(i+1/2)})$ computed here may not fulfil the desired temperature/enthalpy relation, see Fig. 3 (right). Thus, we project the current enthalpy back on the prescribed enthalpy/temperature relation, solving the right hand side of the linearized enthalpy approximation (17) and yielding the temperature update

$$\tilde{T}^{(i+1/2)} = \frac{H^{(i+1/2)}(T_1 - T_0) + T_1 l\Lambda}{c^{(i)}(T_1 - T_0) + l\Lambda}. \tag{19}$$

The freezing state is updated according to the corrected temperature value, $F^{(i+1)} = F^{-1}(\tilde{T}^{(i+1/2)})$ where the invertible relation (6) is used.

Conservation of Energy Special care needs to be taken to maintain the conservation of energy prescribed by the conservation law (1) in the discretized model. Therefore, we introduce a post-processing energy conservation step after the temperature and enthalpy predictor-corrector step.

The state-dependent heat capacity $c(x; T)$ may alter the local energy at a point $x \in \Omega$. Thus, when updating the heat capacity c we must modify T as well to ensure an energy consistent modification of the material based on the enthalpy approximation from Eq. (17). This is done by calculating the "artificial" energy (see Fig. 4) that would result from the heat capacity change if not corrected

$$\tilde{E}(x) = \frac{1}{2} \left(c^{(i+1/2)}(x) - c^{(i)}(x) \right) \left(\tilde{T}^{(i+1/2)}(x) - T^{(i)}(x) \right). \tag{20}$$

Fig. 4 Relation between the artificial energy $\tilde{E}$ and the enthalpy approximation $\tilde{H}$ for the energy-conserving material update

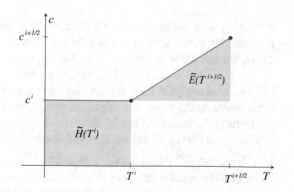

This artificial energy $\tilde{E}$ is then used to correct the local temperature estimate via

$$T^{(i+1)}(x) = \tilde{T}^{(i+1/2)}(x) - \frac{\tilde{E}(x)}{c^{(i+1)}(x)}. \tag{21}$$

Conservation of Mass Mass conservation is ensured by using a constant density ρ, i.e., independent of temperature or freezing state.

Time Stepping Algorithm In summary, we arrive at the following algorithm for the computation of one time-step of the simulation, i.e., computing the values of $T^{(i+1)}$ and $F^{(i+1)}$ given $T^{(i)}$, $F^{(i)}$, and the respective tissue parameters:

1. Compute a predictor $T^{(i+1/2)}$ for the updated temperature through one timestep of the bioheat equation (4) without the enthalpy term according to (17).
2. If the temperature estimate $T^{(i+1/2)}$ lies

   (a) within the mushy region, i.e., $T^{(i+1/2)} \in (T_0, T_1)$, an additional change of the temperature due to the phase change is necessary. In this case, we perform the energy correction step from Sect. 4.1, yielding an updated estimate $\tilde{T}^{(i+1/2)}(x)$.

   (b) outside the mushy region, i.e., $T^{(i+1/2)} \notin (T_0, T_1)$ the energy correction is not applied, thus $\tilde{T}^{(i+1/2)}(x) = T^{(i+1/2)}(x)$.

3. The freezing state is updated according to the corrected temperature value, $F^{(i+1)} = F^{-1}(\tilde{T}^{(i+1/2)})$ where the invertible relation (6) is used.
4. The energy conservation step corrects the temperature $\tilde{T}^{(i+1/2)}$ according to the new value of the heat capacity $c^{(i+1)}$, see Eqs. (20) and (21). This step yields the final temperature update $T^{(i+1)}$.

This algorithm is visualized as a flow diagram in Fig. 5.

4.2 Spatial Discretization

Assuming that the computational domain Ω is always a cuboid, we introduce a regular and isotropic hexahedral grid on Ω with a grid width (resolution) of h. Thus, we introduce nodes $x_{l,m,n} = h(l, m, n)^t \in \Omega$ for $l = 0, \dots, N_x$, $m = 0, \dots, N_y$, $n = 0, \dots, N_z$. Whereas the grid with h is variable from the viewpoint of numerical analysis, for computations on real patient data we choose $h = 0.001$ m comparable to the voxel size of patient-specific image data.

To fully discretize our model, the time-discrete quantities from Sect. 4.1 are only evaluated on these grid points. The corresponding values are denoted by a subscript, i.e., $T_{l,m,n}^{(i)} = T^{(i)}(x_{l,m,n})$, $\rho_{l,m,n}^{(i)} = \rho\left(x; T_{l,m,n}^{(i)}, F_{l,m,n}^{(i)}\right)$, etc. Consequently, we transform the space-continuous time-discrete equations into a fully discrete equation in $(N_x + 1)(N_y + 1)(N_z + 1)$ DOF per state and time step.

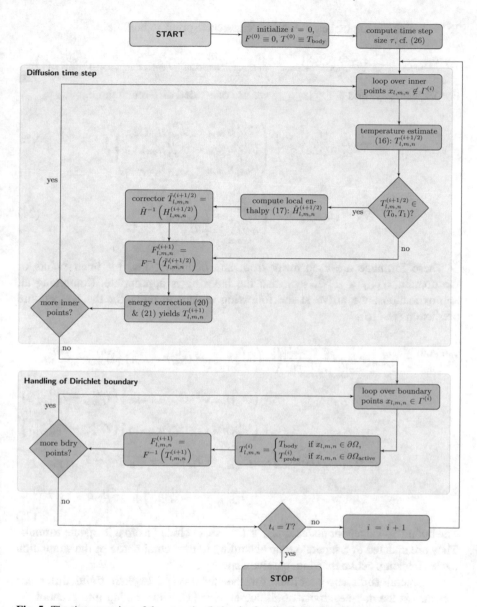

Fig. 5 The time stepping of the cryo simulation is visualized as a flow diagram

To discretize the nabla operator in the predictor equation (16), we use the relation

$$\nabla \cdot \left(k_{\text{eff}}^{(i)}(x) \nabla T^{(i)}(x) \right) = \nabla k_{\text{eff}}^{(i)}(x) \cdot \nabla T^{(i)}(x) + k_{\text{eff}}^{(i)}(x) \Delta T^{(i)}(x). \tag{22}$$

To discretize the Laplacian, we use the classical 7-point stencil, i.e.,

$$\Delta T^{(i)}(x_{l,m,n}) \approx \frac{1}{h^2}\left(6T^{(i)}_{l,m,n} - T^{(i)}_{l-1,m,n} - T^{(i)}_{l+1,m,n} - T^{(i)}_{l,m-1,n} - T^{(i)}_{l,m+1,n} - T^{(i)}_{l,m,n-1} - T^{(i)}_{l,m,n+1}\right).$$
(23)

For $\nabla k^{(i)}_{\text{eff}}(x_{l,m,n})$ and $\nabla T^{(i)}(x_{l,m,n})$ we use one-sided differencing, i.e.,

$$\nabla k^{(i)}_{\text{eff}}(x_{l,m,n}) \approx \frac{1}{h}\begin{pmatrix}(k^{(i)}_{\text{eff}})_{l,m,n} - (k^{(i)}_{\text{eff}})_{l-1,m,n}\\ (k^{(i)}_{\text{eff}})_{l,m,n} - (k^{(i)}_{\text{eff}})_{l,m-1,n}\\ (k^{(i)}_{\text{eff}})_{l,m,n} - (k^{(i)}_{\text{eff}})_{l,m,n-1}\end{pmatrix},$$

$$\nabla T^{(i)}(x_{l,m,n}) \approx \frac{1}{h}\begin{pmatrix}T^{(i)}_{l,m,n} - T^{(i)}_{l-1,m,n}\\ T^{(i)}_{l,m,n} - T^{(i)}_{l,m-1,n}\\ T^{(i)}_{l,m,n} - T^{(i)}_{l,m,n-1}\end{pmatrix},$$
(24)

These formulae are used away from any boundary, i.e., for inner points of the domain $x_{l,m,n} \notin \Gamma^{(i)}$, such that the indexing is appropriate. Combining all approximations, we arrive at the following update formula for the temperature prediction (cf. (16))

$$T^{(i+1/2)}_{l,m,n} = T^{(i)}_{l,m,n} + \frac{\tau}{\rho^{(i)}_{l,m,n}c^{(i)}_{l,m,n}}\frac{1}{h^2}\Bigg(\ \left((k^{(i)}_{\text{eff}})_{l,m,n} - (k^{(i)}_{\text{eff}})_{l-1,m,n}\right)\left(T^{(i)}_{l,m,n} - T^{(i)}_{l-1,m,n}\right)$$

$$+ \left((k^{(i)}_{\text{eff}})_{l,m,n} - (k^{(i)}_{\text{eff}})_{l,m-1,n}\right)\left(T^{(i)}_{l,m,n} - T^{(i)}_{l,m-1,n}\right)$$

$$+ \left((k^{(i)}_{\text{eff}})_{l,m,n} - (k^{(i)}_{\text{eff}})_{l,m,n-1}\right)\left(T^{(i)}_{l,m,n} - T^{(i)}_{l,m,n-1}\right)$$

$$+ (k^{(i)}_{\text{eff}})_{l,m,n}\Big(6T^{(i)}_{l,m,n} - T^{(i)}_{l-1,m,n} - T^{(i)}_{l+1,m,n} - T^{(i)}_{l,m-1,n} - T^{(i)}_{l,m+1,n}$$

$$- T^{(i)}_{l,m,n-1} - T^{(i)}_{l,m,n+1}\Big)\Bigg) \qquad \text{if } x_{l,m,n} \notin \Gamma^{(i)}$$
(25)

The boundary values for nodes $x_{l,m,n} \in \Gamma^{(i)}$ are excluded from this update formula. They are updated in a separate loop according to the actual phase of the simulation cycle (freezing, active thawing, passive thawing).

A general sufficient condition for the stability of explicit finite difference schemes is the non-negativity of all coefficients [1]. Thus, taking into account the above spatial discretization and by estimating the maximum respectively minimum values of the material parameters c, ρ and k_{eff} over all possible temperatures we compute an upper bound $\tilde{\tau}$ for a stable time step as

$$\min_{l,m,n}\left\{\frac{h^2 \min_T c_{l,m,n} \min_T \rho_{l,m,n}}{3\max_T (k_{\text{eff}})_{l,m,n} + \max_T (k_{\text{eff}})_{l-1,m,n} + \max_T (k_{\text{eff}})_{l,m-1,n} + \max_T (k_{\text{eff}})_{l,m,n-1}}\right\} = \tilde{\tau}$$
(26)

The employed time step $\tau \leq \tilde{\tau}$ is chosen to minimize the steps J needed to reach $t_{end} = J\tau$, see Sect. 4.1.

4.3 Discretization of Material Parameters

The spatial discretization with points $x_{l,m,n}$ induces a straightforward discretization of the tissue label function $\mathcal{T}$. Furthermore, all material parameter models from Sect. 3 are discretized by sampling the equations (or tables) at a number temperature values over an interval $[T_{min}, T_{max}]$. We use piecewise linear interpolation between the (not necessarily equidistant) sampling points in $[T_{min}, T_{max}]$ to evaluate any material parameter, with constant extrapolation of the corresponding boundary value outside $[T_{min}, T_{max}]$.

5 Numerical Implementation

The numerical model is discretized in such a way that the computation for all voxels in the domain are independent of each other, i.e., they can be computed in parallel. Therefore, we provide an implementation based on the OpenCL language specifically developed to support *single instruction multiple data* (SIMD) computations on various hardware device, e.g., CPUs as well as GPUs (graphics processing units). OpenCL implementations follow the design of a kernel method (defining the instructions) and a data grid (defining the data to be operated on). Our regular computation domain can be directly used to define the data grids for the OpenCL kernel programs.

Because we use uniform hexahedral grids, the instruments positioned in the domain are rasterized, i.e., the hexahedral grid cells covered by the active zones of the instruments are determined. This set of cells is handled as Dirichlet boundary during freezing and active thawing. The $\partial\Omega$ boundary is implemented by omitting the boundary cells in the explicit updating scheme. This allows always using the full stencils in the update formulas. The condition on $\partial\Omega_{active}$ is implemented as a correction step. After updating the temperature and freezing values in the domain $\Omega \setminus \partial\Omega$, we loop over all voxels in $\partial\Omega_{active}$ and set the respective boundary value dependent on the current simulation cycle state (freezing: T_{needle}, active thaw: T_{body}, passive thaw: skip). The freezing state is set based on the temperature using the mapping F. The implementation utilizes an index set that stores the indices of the nodes that require the correction step.

6 Numerical Examples

Grid Convergence To test the discretization of our model and its implementation, we performed classical grid convergence tests. To this end, we considered the simulation of $t_{end} = 20$ s freezing on a cube grid with $N_x = N_y = N_z = 2^j - 1$, for $j = 3, \ldots, 9$. Thus, we have a total of $(2^j)^3$ degrees of freedom in the domain. We vary the grid with $h = 0.05\,\text{m}/2^j$ accordingly and compare the resulting temperature field $T(t_{end}, \cdot)$ with the result of the temperature simulation on the finest grid $T_{512}(t_{end}, \cdot)$, thus evaluating

$$E[T] = \left\| T(t_{end}, \cdot) - T_{512}(t_{end}, \cdot) \right\|_{L^2(\Omega)}. \tag{27}$$

This numerical experiment was performed for homogeneous tissue (one material present) and for heterogeneous tissue (two different material types present), and the results are reported in Table 7.

Because Ω_{needle} and Ω_{active} are treated as internal boundaries of the domain, we lose the well-known second-order convergence. In our numerical experiments, the grid convergence is between first and second order, see columns 2 and 3 of Table 7. Therefore, we also performed an analysis of the solver of the Pennes bioheat equation (4)/(16) against the well-known analytic solution (convolution with a Gaussian kernel of varying standard deviation) yielding an error $\tilde{E}[T]$. Column 4 of Table 7 shows the expected quadratic convergence. Finally, we analyzed the stability with respect to the time step size τ, which confirmed consistent results compared to the analytic solution, i.e., the L^2 error is in the same order as in Table 7 for the respective grid size.

Energy Conservation To test the freezing model, we evaluated energy conservation of the whole system. We analyzed the energy starting from a frozen state and an

Table 7 We analyze the convergence of our numerical implementation on different grids (first column). We perform a freezing simulation of 20 s with one needle in the domain using homogeneous material (second column) or two different material types (third column). In both cases, we use the solution on the finest grid size (512^3) as ground truth. The last column analyzes the error of the Pennes solver (i.e., the temperature predictor with no needle in the domain) for different grid resolutions using the analytic solution as ground truth

Grid	$E[T]$ Homogeneous	$E[T]$ Heterogeneous	$\tilde{E}[T]$ Analytic
8^3	43.12	43.13	$4.46 \cdot 10^0$
16^3	23.92	24.14	$1.92 \cdot 10^{-1}$
32^3	15.42	15.49	$6.75 \cdot 10^{-3}$
64^3	6.67	6.79	$2.74 \cdot 10^{-5}$
128^3	2.62	2.69	$1.16 \cdot 10^{-5}$
256^3	1.06	1.09	$2.29 \cdot 10^{-7}$
512^3	–	–	$1.48 \cdot 10^{-8}$

Table 8 Performance analysis of a 24-min cryo-cycle of 10 min freezing, 5 min passive thawing, 1 min active thawing, and 8 min freezing with two needles on different grid resolutions. The CPU device is a 4-core Intel(R) i7-6700k @ 4.00 GHz, the GPU device a NVIDIA GTX 980 Ti with 6 GB dedicated device memory. The third column states the acceleration achieved on the GPU compared to the CPU computation time, while the last column shows the acceleration against real time

Grid	Computational time GPU	Computational time GPU	Speedup (GPU vs. CPU)	Speedup (GPU vs. real time)
64^3	2.3 s	28.2 s	12	626
128^3	15.7 s	253.1 s	16	92
256^3	114.5 s	1972.8 s	17	12

unfrozen (body temperature) state, using both homogeneous as well as heterogeneous material to achieve a checkerboard pattern domain. The relative loss of energy was below 0.001 for all of these scenarios.

Performance Finally, we evaluated the runtime of our implementation and compared it to an optimized parallel CPU implementation. In this test, the CPU was a 4-core Intel(R) i7-6700k running at 4.00 GHz, and the GPU was a NVIDIA GTX 980 Ti with 6 GB dedicated device memory. As shown in Table 8, we achieve an acceleration of 12 to 18 times compared to the CPU (depending on the grid size). Even with a grid size of 256^3, we are approximately 12 times faster than real time. The results for the CPU are based on the same optimized OpenCL code.

For clinical applications, we found $h = 0.001$ m (close to standard resolutions used in patient-specific imaging data) to be a good compromise between speed and accuracy. Furthermore, a grid of 128^3 allows for a cuboid simulation domain with 128 mm extent, which is large enough for typical clinical applications.

In comparison to [14], we achieve similar acceleration of about 90 times over real time. However, we can simultaneously handle $128^3 = (2^7)^3 = 2^{21}$ DOFs in a regular grid (with an extent of 128 mm in each dimension), whereas [14] mentions up to 80,000 DOFs in a mixed 1 mm/3 mm regular grid (which allows for a maximal extent of 43 mm in a cubic domain).

Clinical Example Finally, Fig. 6 shows a screenshot of our Software Assistance for Interventional Radiology (SAFIR) [32], which uses the numerical simulation of cryoablation on a real kidney data set. We use different material properties for the tumor, the kidney, and the surrounding tissue. The corresponding domains have been segmented from the patient specific image data. The example shows simulation of freezing for 600 s with a cryoprobe of diameter 1.5 mm and cooling temperature of 113.15 K.

As mentioned before, for the patient specific image data we use a resolution of $h = 0.001$ m and a hexahedral grid of 128^3 DOFs. In the vicinity of the probe, a part of the segmented image data with this size is passed on to the cryo-simulation in order to define the tissue types according to Sect. 3. Figure 7 shows the location of the computational domain within the 3D patient data set. Also, Fig. 7 shows

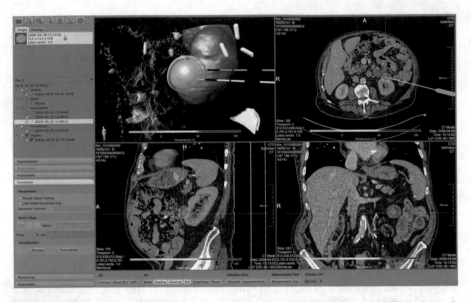

Fig. 6 A screenshot of our Software Assistance for Interventional Radiology (SAFIR) is shown. It visualizes the simulation of cryoablation in a real kidney dataset overlaid on patient-specific image data. The kidney and a tumor are segmented in orange. Isosurfaces/isolines show the temperature field obtained from the simulation. (Dataset courtesy RWTH Aachen)

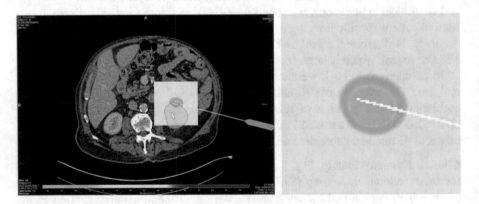

Fig. 7 For the application to patient specific data that is depicted in Fig. 6 we show the computational domain for the temperature computation. A cuboid domain with hexahedral grid of extent 128^3 mm, corresponding to 128^3 DOF, is used to compute the temperature in the target region. *Left*: The location of the computational grid within the 3D patient image data set is shown as one slice through the 3D volume. *Right*: The temperature is shown on one slice of the computational domain. The rasterization of the cryoprobe, on which the temperature boundary conditions are prescribed, is visible at white voxels/pixels. In both images, color codes the temperature according to the scale shown in Fig. 6

the computed temperature on one slice of the 128^3 computational grid in which the rasterization of a cryoprobe is visible. On these voxels the boundary conditions from (14) are set.

7 Summary, Conclusions, and Future Work

We have presented a model for cryoablation tailored towards implementation on graphics hardware. We use explicit time-stepping and a predictor corrector approach to account for the non-linearities resulting from the phase change of liquid to frozen water. Values for material parameters have been taken from literature.

To test our solver we have performed classical convergence analysis tests for both the full model and the Pennes bioheat equation. For the full model we obtain convergence rates between 1 and 2, which is to be expected because of the boundary conditions on the cryoprobes that are set in the interior of the domain. Our Pennes bioheat equation solver has been tested against an analytical solution and yields convergence rates above 2.

We have furthermore applied our simulation of cryoablation to real patient data in which we use a resolution of $h = 0.001$ m and 128^3 DOFs. In such situation our algorithm needs about 17 seconds to compute the tissue temperature at the end of a typical cryoablation cycle. Compared to our CPU implementation this is a speedup of 16 times. Such performance seems to be suitable for applications in the daily clinical routine.

As outlined in the introduction, our aim is to include the presented algorithm into a medical product, for which a rigorous validation, risk analysis, and sensitivity analysis must be performed. Reporting our activities in this regard is ongoing and future work. We plan to compare our temperature simulations to data obtained on phantoms and real patient data sets. Also, we will report how the results change with variations of parameters, and finally we will investigate wether the results of the temperature simulation can be improved through parameter variations of the tissue parameter models that we considered in Sect. 3.

Acknowledgements We acknowledge Prof. A. Mahnken for providing the dataset shown in Fig. 6. Furthermore, we thank our colleague D. Black for proofreading the manuscript. This work was partly supported by the FhG Internal Programs under Grant No. MEF 142-600369.

References

1. H.D. Baehr, K. Stephan, *Wärme- und Stoffübertragung* (Springer, Heidelberg, 2010)
2. R. Baissalov, G.A. Sandison, B.J. Donnelly, J.C. Saliken, J.G. McKinnon, K. Muldrew, J. C. Rewcastle, A semi-empirical treatment planning model for optimization of multiprobe cryosurgery. Phys. Med. Biol. **45**(5), 1085–1098 (2000)

3. H.M. Cheng, D.B. Plewes, Tissue thermal conductivity by magnetic resonance thermometry and focused ultrasound heating. J. Magn. Reson. Imaging **16**, 598–609 (2002)

4. J.H. Choi, J.C. Bischof, Review of biomaterial thermal property measurements in the cryogenic regime and their use for prediction of equilibrium and non-equilibrium freezing applications in cryobiology. Cryobiology **60**(1), 52–70 (2010)

5. R. Courant, K.O. Friedrichs, H. Lewy, On the partial difference equations of mathematical physics. IBM J. **11**(2), 215–234 (1967). https://doi.org/10.1147/rd.112.0215

6. G. Deshazer, D. Merck, M. Hagmann, D. Dupuy, P. Prakash, Physical modeling of microwave ablation zone clinical margin variance. Med. Phys. **43**(4), 1764–1776 (2016)

7. G. Deshazer, P. Prakash, D. Merck, D. Haemmerich, Experimental measurement of microwave ablation heating pattern and comparison to computer simulations. Int. J. Hyperthermia **33**(1), 74–82 (2017)

8. D.K. Filippiadis, S. Tutton, A. Mazioti, A. Kelekis, Percutaneous image-guided ablation of bone and soft tissue tumours: a review of available techniques and protective measures. Insights Imaging **5**(3), 339–346 (2014)

9. A.A. Gage, J. Baust, Mechanisms of tissue injury in cryosurgery. Cryobiology **37**(3), 171–186 (1998)

10. L. Gavrilă, A. Fînaru, L. Istrati, A.I. Simion, M.E. Ciocan, Influence of temperature, fat and water content on the thermal conductivity of some dairy products. Sci. Res. J. Agroaliment. Processes Technol. **11**(1), 205–210 (2005)

11. P.A. Hasgall, F. Di Gennaro, C. Baumgartner, E. Neufeld, B. Lloyd, M.C. Gosselin, D. Payne, A. Klingenboeck, N. Kuster, IT'IS database for thermal and electromagnetic parameters of biological tissues, version 4.0 (2018)

12. L.J. Hayes, K.R. Diller, Implementation of phase change in numerical models of heat transfer. J. Energy Resour. Technol. **105**(4), 431–435 (1983)

13. International Organization for Standardization, Medical devices – quality management systems – requirements for regulatory purposes (ISO Norm 13485:2016) (2016)

14. R. Keelan, H. Zhang, K. Shimada, Y. Rabin, Graphics processing unit-based bioheat simulation to facilitate rapid decision making associated with cryosurgery training. Technol. Cancer Res. Treat. **15**(2), 377–386 (2016)

15. F. Liu, P. Liang, X. Yu, T. Lu, Z. Cheng, C. Lei, Z. Han, A three-dimensional visualisation preoperative treatment planning system in microwave ablation for liver cancer: a preliminary clinical application. Int. J. Hyperthermia **29**(7), 671–677 (2013)

16. H.H. Mitchell, T.S. Hamilton, F.R. Steggerda, H.W. Bean, The chemical composition of the adult human body and its bearing on the biochemistry of growth. J. Biol. Chem. **158**, 625–637 (1945)

17. H.H. Pennes, Analysis of tissue and arterial blood temperatures in the resting human forearm. J. Appl. Physiol. **1**(2), 93–122 (1948)

18. Q.T. Pham, J. Willix, Thermal conductivity of fresh lamb meat, offals and fat in the range −40 to +30 °C: measurements and correlations. J. Food Sci. **54**(3), 508–515 (1989)

19. Y. Rabin, K. Shimada, P. Joshi, A. Sehrawat, R. Keelan, D.M. Wilfong, J.T. McCormick, A computerized tutor prototype for prostate cryotherapy: key building blocks and system evaluation. Proc. SPIE Int. Soc. Opt. Eng. 10066 (2017)

20. Y. Rabin, A. Shitzer, Numerical solution of the multidimensional freezing problem during cryosurgery. J. Biomech. Eng. **120**(1), 32–37 (2007)

21. K.F. Ross, R.E. Gordon, Water in malignant tissue, measured by cell refractometry and nuclear magnetic resonance. J. Microsc. **128**(1), 7–21 (1982)

22. M.R. Rossi, Y. Rabin, Experimental verification of numerical simulations of cryosurgery with application to computerized planning. Phys. Med. Biol. **52**(15), 4553–4567 (2007)

23. M.R. Rossi, D. Tanaka, K. Shimada, Y. Rabin, An efficient numerical technique for bioheat simulations and its application to computerized cryosurgery planning. Comput. Methods Programs Biomed. **1**(85), 41–50 (2007)

24. C. Rossmann, D. Haemmerich, Review of temperature dependence of thermal properties, dielectric properties, and perfusion of biological tissues at hyperthermic and ablation temperatures. Crit. Rev. Biomed. Eng. **42**(6), 467–492 (2014)
25. Rotes Kreuz Austria, Die Zusammensetzung unseres Blutes, call: 2018/09/19. https://www.roteskreuz.at/blutspende/blut-im-detail/wissenswertes-ueber-blut/blutbestandteile/
26. R. Schmidt, G. Thews, *Physiologie des Menschen* (Springer, Berlin/Heidelberg, 1995)
27. C. Schumann, C. Rieder, J. Bieberstein, A. Weihusen, S. Zidowitz, J.H. Moltz, T. Preusser, State of the art in computer-assisted planning, intervention, and assessment of liver-tumor ablation. Crit. Rev. Biomed. Eng. **38**(1), 31–52 (2010)
28. A. Sehrawat, R. Keelan, K. Shimada, D.M. Wilfong, J.T. McCormick, Y. Rabin, Simulation-based cryosurgery training: variable insertion depth planning in prostate cryosurgery. Technol. Cancer Res. Treat. **15**(6), 805–814, 12 (2016)
29. T.-C. Shih, H.-S. Kou, W.-L. Lin, Effect of effective tissue conductivity on thermal dose distributions of living tissue with directional blood flow during thermal therapy. Int. Commun. Heat Mass Transf. **29**(1), 115–126 (2002)
30. T. Stein, *Untersuchungen zur Dosimetrie der hochfrequenzstrominduzierten interstitiellen Thermotherapie in bipolarer Technik*, vol. 22 of *Fortschritte in der Lasermedizin* (Ecomed Verlagsgesellschaft AG & Co. KG, 2000)
31. J.W. Valvano, J.R. Cochran, K.R. Diller, Thermal conductivity and diffusivity of biomaterials measured with self-heated thermistors. Int. J. Thermophys. **6**(3), 301–311 (1985)
32. A. Weihusen, F. Ritter, T. Kröger, T. Preusser, S. Zidowitz, H.-O. Peitgen, in *Workflow Oriented Software Support for Image Guided Radiofrequency Ablation of Focal Liver Malignancies*, ed. by K.R. Cleary, M.I. Miga, Proceedings SPIE (2007), pp. 650919–650919–9
33. S. Weinbaum, L.M. Jiji, A new simplified bioheat equation for the effect of blood flow on local average tissue temperature. J. Biomech. Eng. **107**(2), 131–139 (1985)
34. S. Yılmaz, M. Özdoğan, M. Cevener, A. Ozluk, A. Kargi, F. Kendiroglu, I. Ogretmen, A. Yildiz, Use of cryoablation beyond the prostate. Insights Imaging **7**(2), 223–232 (2016)
35. J. Zhang, G.A. Sandison, J.Y. Murthy, L.X. Xu, Numerical simulation for heat transfer in prostate cancer cryosurgery. J. Biomech. Eng. **127**(2), 279–294, 09 (2004)

Index

A

AAA, *see* Abdominal aortic aneurysm (AAA)
Abdominal aortic aneurysm (AAA), 111, 114, 148
Ad-hoc evolutionary algorithm, 6
Aforementioned ablation techniques, 172
Anisotropic electrical conductivity, 138
Antiepileptic drugs
    FCFD method, 137
    iEEG/sEEG, 136
    mesh-based methods, 136
    patient-specific head model, 136
    SOZ, 136
    treatment strategy, 135
Atlas-based 3D shape reconstruction, 15
Auxetic SE (ASE) stents, 19
    mechanic requirements, 29
    numerical results (*see* Numerical results, ASE)

B

Balloon-expandable (BE), 18
BCM, *see* Bilayer-coupling model (BCM)
BCPM, *see* Bone cell population models (BCPM)
BE, *see* Balloon-expandable (BE)
Bilayer-coupling model (BCM), 58
BIM, *see* Boundary integral method (BIM)
Bioelectric field, 142
Bioresorbable vascular scaffolds (BVS), 18
BMD, *see* Bone mineral density (BMD)
BMU, *see* Bone multicellular units (BMU)
Bone cell population models (BCPM), 86
Bone mineral density (BMD), 86

Bone multicellular units (BMU), 93
Bone turnover markers (BTMs), 86
Bone volume fraction (BV/TV), 86
Boundary integral method (BIM), 54
Breast biopsy
    image guidance, 34
    suspicious lesions, 34
    ultrasound (US) (*see* US-guided breast biopsy)
BTMs, *see* Bone turnover markers (BTMs)
BVS, *see* Bioresorbable vascular scaffolds (BVS)
BV/TV, *see* Bone volume fraction (BV/TV)

C

CAD, *see* Coronary artery disease (CAD)
Cerebrospinal fluid (CSF), 1, 143
CFL, *see* Courant-Friedrichs-Lewy (CFL)
CGMD, *see* Coarse-grained molecular dynamics (CGMD)
CG-RBC membrane model
    computational implementation, 67
    cytoskeletal reference state, 66
    development, 59–60
    discocyte morphology
        evolution of shape, 69
        parameters, 67–69
        quantitative validation, 77
        variation, 69–72
    free-energy, 60–62
    spherical geometry, 62–64
        minimum particle resolution, 65–66
        triangulation quality, 64–65
Chiari malformation, 1

© Springer Nature Switzerland AG 2020
K. Miller et al. (eds.), *Computational Biomechanics for Medicine*,
https://doi.org/10.1007/978-3-030-42428-2

Clinical target volume (CTV), 6, 15
Coarse-grained molecular dynamics (CGMD),
        54
Computer assisted EVAR, *see* Endovascular
        surgery (EVAR)
Concurrent topology optimization scheme
    bilinear energy, 23
    CSRBF interpolation, 23
    finite element model, 23
    macrostructure, 22
    microstructure, 22
    multi-objective optimization problem, 22
Continuum-based modelling, 54
Coronary artery disease (CAD)
    ASE stents, 19
    BDS, 18
    BE stents, 18
    BMS, 18
    IHD, 17
    level set function, 20
    NPR, 19
    self-expanding procedure, 19
    topology optimization, 19 (*see* Topology
        optimization)
Courant-Friedrichs-Lewy (CFL), 19, 181
Cryoablation
    damage model, 176–177
    freezing model, 176–177
    temperature model, 174–176
CSF, *see* Cerebrospinal fluid (CSF)
CT, *see* Computed tomography (CT)
CTV, *see* Clinical target volume (CTV)
Cytoplasmic viscosity, 52

D
DES, *see* Drug-eluting stents (DES)
Discretization
    spatial, 184–187
    temporal, 181–184
Dissipative particle dynamics (DPD), 54
DPD, *see* Dissipative particle dynamics (DPD)
Drug-eluting stents (DES), 18

E
Echinocyte, 51
Endovascular surgery (EVAR)
    barycenter positioning, 154–156
    definition, 148
    FVS, 149
    geometric stent reconstruction, 156–158
    individual stent deployment, 158–160
    methods

    corotational Euler-Bernoulli Beam
        Elements, 151–152
    data acquisition, 150–151
    FEM model, 149
    3D positioning, algorithm, 152–153
    proof of concept
        clinical data, 161
        quality assessment, 161–162
        results, 163–166
    rotational difference, 159
    SG, 148
    X-rays, 148
Epilepsy, 135, 137
ESO, *see* Evolutionary structural optimization
        (ESO) method
EVAR, *see* Endovascular surgery (EVAR)
Evolutionary structural optimization (ESO)
        method, 19

F
Fast Virtual Stenting (FVS) technique, 149
FEM, *see* Finite element method (FEM)
FENE, *see* Finitely extensible non-linear
        elastic (FENE) models
Finite element method (FEM), 6, 10, 34, 54,
        136
Finitely extensible non-linear elastic (FENE)
        models, 57
Flux-conservative finite difference (FCFD)
    antiepileptic drugs, 135
    electric potential distribution, 145
    electromagnetic modeling, 137–138
    epilepsy, 135
    FD method, 138–141
    point-electrode model, 144
    source localization, 136
    verification, 142–143
Flux-conservative finite difference (FCFD)
        method
    anisotropic tensor, 137
    Cartesian grids, 138
    EEG source localization, 137
    electric potential distribution, 144, 145
    epileptogenic cortex, 144
    inhomogeneous anisotropic medium,
        141–143
    interpolating/approximating methods,
        139
    Laplace equation, 142
    linear model/leadfield matrix, 137–138
    NRMSE, 142
    off-grid nodes, 139–140
    patient-specific head model, 143–144

point-electrode model, 144
   3D stencil configuration, 139, 141
Freehand Ultrasound System (FUS), 39, 40,
   172

G
Glucocorticoid-induced osteoporosis, 98
GPU-based numerical simulation
   classical convergence analysis tests, 191
   clinical applicability, 172
   cryoablation, 172, 173
   enthalpy approach, 174, 177, 178
   examples, 188–191
   implementation, 187
Green-Lagrange strain tensor, 9

H
Hamilton-Jacobi Partial Differential Equation
   (PDE), 19–21
High intensity focused ultrasound (HIFU), 172
Hyper-thermal ablation, 171

I
IBM, *see* Immersed boundary method (IBM)
IHD, *see* Ischemic heart disease (IHD)
Immersed boundary method (IBM), 54
In-stent thrombosis (ISR), 18, 19, 25
Internal margin (IM), 6
Invasive electroencephalography (iEEG) grids,
   136
Invasive electroencephalography
   (iEEG)/stereo-EEG (sEEG),
   136
Ischemic heart disease (IHD), 17
ISR, *see* In-stent thrombosis (ISR)

L
Lagrange surface mesh, 129
Laser light (LITT), 172
Lattice Boltzmann method (LBM), 54
LBM, *see* Lattice Boltzmann method (LBM)
Leadfield matrix, 137
Left-posterior (RP), 11
Level set method (LSM), 19
Linear model/leadfield matrix, 137–138
Lipoprotein receptor-related protein (LRP), 94
LRP, *see* Lipoprotein receptor-related protein
   (LRP)
LSM, *see* Level set method (LSM)

Lungs
   computational modelling, 124
   image registration method, 131
   imaging technologies, 124
   intrinsic features, 130–131
   micro-CT imaging, 128–129
   model-based reconstruction approach, 132,
      133
   pipeline, construction, 133
   quadratic Lagrange mesh, 133
   respiratory mechanics, 124
   stereoscopic imaging, 124, 134
   surface imaging (*see* Surface imaging,
      lungs)
   surface reconstruction and tracking motion,
      129–130
   3D reconstruction, 131–132
   3D tracking, 134
   ventilation
      cannulated rat, 125
      PV loops, 125
      Sprague-Dawley strain, 124
Lung tumor tracking
   automatic tuning algorithm, 6
   biomechanical approaches, 6
   correspondence models, 6
   4D CT images, 6
   IM, 6
   patient-specific information, 6
   physiological pressure-volume curve, 6
   respiratory motion, 6
   SM, 6

M
Machine learning-based methods, 35
Magnetic resonance images (MRIs), 136
MD, *see* Molecular dynamics (MD)
Mesh quality
   Abaqus packages, 11
   breathing cycle and intermediate states, 13
   displacement field of lungs, 11, 13
   FE lung tumor, 12
   high stress concentration, 11
   tetrahedral elements, 12
   3D lung tumor trajectories, 12, 14
Micro Computed Tomography (CT) imaging
   air-drying process, 129
   glutaraldehyde, 128
   lung lobe, 129
Microwaves (MWA), 172
Molecular dynamics (MD), 54
MRIs, *see* Magnetic resonance images (MRIs)

N

Negative Poisson's ratios (NPR), 19, 22
Newton-Raphson method, 38
Non-uniform memory access (NUMA), 55
Normalized root mean square error (NRMSE),
    142
NPR, *see* Negative Poisson's ratios (NPR)
NRMSE, *see* Normalized root mean square
    error (NRMSE)
NUMA, *see* Non-uniform memory access
    (NUMA)
Numerical homogenization method, 19
    heaviside function, 21
    periodical boundary conditions, 22
    2D microstructure, 21
Numerical results, ASE
    macro structure, 27
    microstructures, 26
    NPR properties, 25
    optimized structure, 27
    Poisson's ratios, 25
    stent structure, 27
NURBS, *see* Non-uniform rational B-spline
    (NURBS) curves

O

Obp, *see* Osteoblast precursor cells (Obp)
OP, *see* Osteoporosis (OP)
Osteoblast precursor cells (Obp), 89
Osteoclasts, 100
Osteoporosis (OP), 86
    anabolic/catabolic responses
        mechanically-controlled signalling
            pathways, PMO, 96–97
        mechanobiological feedback, 98
    anti-sclerostin antibody therapy, 94–95
    biochemical mechanostat feedback, 88
    biochemical parameters, 103
    mechanistic tissue-scale model of bone
        remodeling, 88–90
    osteocyte-driven mechanical feedback
        action of nitric oxide, 93
        mechanobiological feedback loop, 93
        mechanostat feedback, 90–91
        sclerostin, 92–93
    osteoporosis-induced bone loss, 98–99
    PK-PD, 86
    post-menopausal, 94
    virtual anti-sclerostin therapy,
        99–100

P

Parametric level set method (PLSM), 19
    Hamilton-Jacobi PDE, 20, 21
    parametric form, 21
    zero level set, 20
Patient-specific head model, 143–144
PBD, *see* Position-based dynamics (PBD)
    approach
PBS, *see* Phosphate buffered saline (PBS)
PCI, *see* Percutaneous coronary intervention
    (PCI)
Percutaneous coronary intervention (PCI), 17
Pharmacokinetic (PK) models, 95, 102
Pharmacokinetic-pharmacodynamic (PK-PD)
    modeling, 86
Phosphate buffered saline (PBS), 126
Piola-Kirchhoff stress tensor, 9
PK, *see* Pharmacokinetic (PK) models
PK-PD, *see* Pharmacokinetic-
    pharmacodynamic (PK-PD)
    modeling
Planning target volume (PTV), 6, 15
PLSM, *see* Parametric level set method
    (PLSM)
PMO, *see* Post-menopausal osteoporosis
    (PMO)
PMO-induced bone loss, 101
POD, *see* Proper Orthogonal Decomposition
    (POD)
Position-based dynamics (PBD) approach, 35
Post-menopausal osteoporosis (PMO), 99
Pressure-volume (PV) loops, 125
Proper Orthogonal Decomposition (POD), 34
PTV, *see* Planning target volume (PTV)

R

Radio-frequency current (RFA), 172
RANK-RANKL-OPG pathway, 101
RBC, *see* Red blood cells (RBCs)
RBC-RBC interactions, 55
Red blood cells (RBCs)
    BCM, 58
    biconcave shape, 56
    cellular structure, 49–50
    CG-RBC membrane model *vs.* deformation
        behaviour of discocyte cell, 76
    Coarse-graining (CG) (*see* CG-RBC
        membrane model)
    composition, 48
    cross-section view, 49

deformability, 51–53
FENE, 57
in-plane stretching energy, 56
measurement, 75
morphologies, 50–51
numerical investigations, 53–55
numerical predictions, 55–59
optical tweezers stretching
    implementation, 73–74
    validation, 74–75
out-of-plane bending energy, 58
SCM, 58
SEM images, 51
stomatocyte/echinocyte, 51
Region of interest (ROI), 131
Respiratory system
    anatomy/physiology, 7
    biomechanical patient-specific model, 8–9
    boundary conditions (BC), 10
    3D segmentation/CAD reconstruction, 7–8
Right-posterior (RP), 11
RP, see Left-posterior (RP)

S
SAGM, see Saline-adenine-glucose-mannitol
        (SAGM)
Saint Venant-Kirchhoff model, 38
Saline-adenine-glucose-mannitol (SAGM), 51
Sclerostin, 92–93
SCM, see Spontaneous curvature model (SCM)
SE, see Self-expanding (SE)
Second topology optimization, numerical
        validation
    auxetic property, 28
    commercial software ANSYS v19.2, 27
    result, 28
    stretching test, 28
SED, see Strain energy density (SED)
Seizure onset zone (SOZ), 136
Sensitivity analysis
    design variables, 25
    effective elasticity tensor, 24
    first-order derivatives, 24
Setup margin (SM), 6
Sickle-cell anaemia, 78
SIMP, see Solid isotropic material with
        penalization (SIMP) method
Simulation of breast tissue, hexahedral grids
    Cauchy stress tensor, 38
    Dirichlet boundary conditions, 37

FE method, 37
    Neumann boundary conditions, 37
    pre-operative CT scan, 38
Solid isotropic material with penalization
        (SIMP) method, 19
SOZ, see Seizure onset zone (SOZ)
Spontaneous curvature model (SCM), 58
Stent graft (SG), 148
Stent thrombosis (ST), 18
stereo-EEG (sEEG) electrodes, 136
Stomatocyte, 51
Strain energy density (SED), 87
Surface imaging, lungs
    camera calibration, 126–127
    initial surface reconstruction, 127–128
    pressure-controlled inflation, 126
    stereo rig construction, 126
    3D reconstruction, 126

T
Target registration error (TRE), 41
Tissue parameter model
    specific heat capacity, 179–181
    thermal conductivity, 178, 179
Topology optimization, 19
    concurrent (see Concurrent topology
        optimization scheme)
    PLSM (see Parametric level set method
        (PLSM))
Total Lagrangian explicit dynamics (TLED),
        34
TRE, see Target registration error (TRE)

U
U-net architecture
    decoding path, 36
    encoding path, 36
    padded input grid, 37
    volumetric displacement field, 36
US-guided breast biopsy
    deep neural network, 43
    displacement field, 40
    FEM, 34
    FE model, 43
    FE simulations, 35
    immersed boundary simulations, 43
    Landmark-based rigid registration, 39
    maximal nodal deformation, 41
    mean norm error, 40

US-guided breast biopsy (*cont.*)
   meshing process, 43
   neural network, 35
   patient-specific model, 35
   random probe-induced deformations, 38
   simulation, Hexahedral grids, 37–38
   soft tissue deformation, 34
   TLED, 34
   TRE, 42–43

2D freehand US (FUS) systems, 34
   U-Net architecture, 35
US probe
   rigid body, 36
   surface nodes, 36

V
Virtual anti-sclerostin therapy, 99–100

Printed in the United States
by Baker & Taylor Publisher Services